T0212537

Lecture Notes in Computer Science   10138

Commenced Publication in 1973
Founding and Former Series Editors:
Gerhard Goos, Juris Hartmanis, and Jan van Leeuwen

More information about this series at http://www.springer.com/series/7407

Klaus Jansen · Monaldo Mastrolilli (Eds.)

Approximation and Online Algorithms

14th International Workshop, WAOA 2016
Aarhus, Denmark, August 25–26, 2016
Revised Selected Papers

 Springer

Editors
Klaus Jansen
Institut für Informatik
Christian-Albrechts-Universität
Kiel
Germany

Monaldo Mastrolilli
Istituto Dalle Molle di Studi
    sull' Intelligenza Artificiale
Manno (Lugano)
Switzerland

ISSN 0302-9743   ISSN 1611-3349   (electronic)
Lecture Notes in Computer Science
ISBN 978-3-319-51740-7   ISBN 978-3-319-51741-4   (eBook)
DOI 10.1007/978-3-319-51741-4

Library of Congress Control Number: 2016961300

LNCS Sublibrary: SL1 – Theoretical Computer Science and General Issues

Printed on acid-free paper

This Springer imprint is published by Springer Nature
The registered company is Springer International Publishing AG
The registered company address is: Gewerbestrasse 11, 6330 Cham, Switzerland

Preface

This volume contains the revised selected papers presented at WAOA 2016: the 14th Workshop on Approximation and Online Algorithms held during August 25–26, 2016, in Aarhus. WAOA 2016 focused on the design and analysis of approximation and online algorithms. These algorithms have become a fundamental tool in several fields and in many applications that cope with computationally hard problems and problems in which the input is gradually disclosed over time.

WAOA 2016 was part of ALGO 2016, which also hosted ESA, ALGOCLOUD, ALGOSENSORS, ATMOS, IPEC, and MASSIVE. The previous WAOA workshops were held in Budapest (2003), Rome (2004), Palma de Mallorca (2005), Zurich (2006), Eilat (2007), Karlsruhe (2008), Copenhagen (2009), Liverpool (2010), Saarbrücken (2011), Ljubljana (2012), Sophia Antipolis (2013), Wraclaw (2014), and Patras (2015). The proceedings of all these previous WAOA workshops have been published as LNCS volumes.

Topics of interest for WAOA 2016 were: coloring and partitioning, competitive analysis, network design, packing and covering, paradigms for design and analysis of approximation and online algorithms, randomization techniques, real-world applications, and scheduling problems.

In response to the call for papers, we received 33 submissions. Each submission was reviewed by at least three referees, and mainly judged on originality, technical quality, and relevance to the topics of the conference. Based on the reviews, the Program Committee selected 16 papers. This volume contains final revised versions of these papers. In addition to the accepted contributions, the workshop featured two invited lectures by Marek Cygan (University of Warsaw, Poland) and Ronald de Wolf (CWI and University of Amsterdam, The Netherlands). Contributions of the invited lectures are also included in this volume. We are grateful to both of them for accepting our invitation and for their very nice lectures.

The EasyChair conference system was used to manage the electronic submissions, the review process, and the electronic Program Committee meeting. It made our task much easier. We wish to thank all the authors who submitted papers for consideration, the invited speakers, the members of the Program Committee for their work, and all the external reviewers who assisted the Program Committee in the evaluation process. Special thanks go to the local Organizing Committee, who helped us with the organization of the workshop.

November 2016

Klaus Jansen
Monaldo Mastrolilli

Organization

Program Committee

Per Austrin	KTH Royal Institute of Technology, Sweden
Nikhil Bansal	Eindhoven University of Technology, The Netherlands
Jose Correa	Universidad de Chile, Chile
Marek Cygan	University of Warsaw, Poland
Michael Fellows	University of Bergen, Norway
Samuel Fiorini	Université Libre de Bruxelles, Belgium
Naveen Garg	IIT Delhi, India
Fabrizio Grandoni	IDSIA, Switzerland
Luciano Gualà	University of Rome Tor Vergata, Italy
Klaus Jansen	University of Kiel, Germany
Jochen Koenemann	University of Waterloo, Canada
Monaldo Mastrolilli	IDSIA, Switzerland
Nicole Megow	Technische Universität München, Germany
Benjamin Moseley	Washington University, USA
Vangelis Paschos	University of Paris-Dauphine, France
Andreas S. Schulz	Technische Universität München, Germany
Roberto Solis-Oba	The University of Western Ontario, Canada
Leen Stougie	Vrije Universiteit and CWI Amsterdam, The Netherlands
Ola Svensson	EPFL, Switzerland
Rob van Stee	University of Leicester, UK

Additional Reviewers

Bilò, Davide	Kumar, Nikhil	Page, Daniel R.
Chen, Lin	Kurpisz, Adam	Paulsen, Niklas
de Keijzer, Bart	Laekhanukit, Bundit	Rau, Malin
Doerr, Benjamin	Lampis, Michael	Sitters, Rene
Fotakis, Dimitris	Leppänen, Samuli	Stamoulis, Georgios
Khan, Arindam	Maack, Marten	Udwani, Rajan
Krumke, Sven	Matuschke, Jannik	van Ee, Martijn
Kumar, Amit	Murat, Cécile	Verschae, José

Invited Lectures

Approximation Algorithms
for the k-Set Packing Problem

Marek Cygan

Approximation Algorithms
for the k-Set Packing Problem

Marek Cygan

Institute of Informatics, University of Warsaw, Warsaw, Poland
`cygan@mimuw.edu.pl`

Abstract. In the k-Set Packing problem we are given a universe and a family of its subsets, where each of the subsets has size at most k. The goal is to select a maximum number of sets from the family which are pairwise disjoint. It is a well known NP-hard problem, that has been studied from the approximation perspective since the 80's. During the talk we describe the history of progress on both the weighted and un- weighted variants of the problem, with an exposition of methods used to obtain the best known approximation algorithms mostly involving local search based routines.

We start with an exemplatory example of the classic Maximum Matching problem. Even though this is a polynomial-time-solvable problem it serves well at explaining the intuition behind local search algorithm in the form of hill climbing. In particular we will see that if it impossible to improve a matching M by removing p and adding $p + 1$ edges, then M is at most $(p + 2)/(p + 1)$ times smaller than optimum.

Next we move to the k-Set Packing problem and consider the canonical local search algorithm for this problem, the approximation ratio of which has been analyzed in a long-spanning sequence of papers [5, 6, 7]. For each of the mentioned results we underline its main idea.

As the standard local search provides better approximation ratio in quasi-polynomial time than in polynomial time, a natural direction was to explore the logarithmic radius search space in polynomial time. This was achieved by Sviridenko and Ward [8] and Cygan [4] by using tools from parameterized complexity such as color coding of Alon, Yuster and Zwick [1].

Even though the standard linear relaxation of the problem has integrality gap $k - 1 + 1/k$ it was shown by Chan and Lau [3] that by adding clique constraints the gap may be upper bounded by $(k + 1)/2$.

Finally we consider the weighed variant of the k-Set Packing problem, where the interesting aspect is that the best known approximation algorithm is a local search optimizing the sum of squares of weights instead of the standard weighted sum [2].

References

1. Alon, N., Yuster, R., Zwick, U.: Color-coding. J. ACM **42**(4), 844–856 (1995)
2. Berman, P.: A $d/2$ approximation for maximum weight independent set in d-claw free graphs. In: Halldórsson, M.M. (ed.) SWAT 2000. LNCS, vol. 1851, pp. 214–219. Springer, Heidelberg (2000)

3. Chan, Y.H., Lau, L.C.: On linear and semidefinite programming relaxations for hypergraph matching. Math. Program. **135**(1–2), 123–148 (2012)
4. Cygan, M.: Improved approximation for 3-dimensional matching via bounded path-width local search. In: Proceedings of FOCS 2013, pp. 509–518 (2013)
5. Cygan, M., Grandoni, F., Mastrolilli, M.: How to sell hyperedges: the hypermatching assignment problem. In: Proceedings of SODA 2013, pp. 342–351 (2013)
6. Halldórsson, M.M.: Approximating discrete collections via local improvements. In: Proceedings of SODA 1995, pp. 160–169 (1995)
7. Hurkens, C.A.J., Schrijver, A.: On the size of systems of sets every t of which have an SDR, with an application to the worst-case ratio of heuristics for packing problems. SIAM J. Discrete Math. **2**(1), 68–72 (1989)
8. Sviridenko, M., Ward, J.: Large neighborhood local search for the maximum set packing problem. In: Fomin, F.V., Freivalds, R., Kwiatkowska, M., Peleg, D. (eds.) ICALP 2013. LNCS, vol. 7965, pp. 792–803. Springer, Heidelberg (2013)

On Linear and Semidefinite Programs for Polytopes in Combinatorial Optimization

Ronald de Wolf

CWI and University of Amsterdam, Amsterdam, The Netherlands
rdewolf@cwi.nl

Ronald de Wolf—Partially supported by ERC Consolidator Grant QPROGRESS

Abstract. Combinatorial problems like TSP optimize a linear function over some polytope P. If we can obtain P as a projection from a larger-dimensional polytope with a small number of facets, then we get a small linear program for the optimization problem; if we obtain P as a projection from a small spectrahedron, then we get a small semidefinite program. The area of extension complexity studies the minimum sizes of such LPs and SDPs. In the 1980s Yannakakis [7] was the first to do this, proving exponential lower bounds on the size of symmetric LPs for the TSP and matching polytopes. In 2012, Fiorini et al. [4] proved exponential lower bounds on the size of all (possibly non-symmetric) LPs for TSP. This was followed by many new results for LPs and SDPs, for exact optimization as well as for approximation. We will survey this recent line of work [1, 2, 3, 5, 6].

References

1. Braun, G., Fiorini, S., Pokutta, S., Steurer, D.: Approximation limits of linear programs (beyond hierarchies). In: Proceedings of 53rd IEEE FOCS, pp. 480–489 (2012). arXiv:1204.0957
2. Braverman, M., Moitra. A.: An information complexity approach to extended formulations. In: Proceedings of 45th ACM STOC, pp. 161–170, (2013)
3. Chan, S.O., Lee, J.R., Raghavendra, P., Steurer, D.: Approximate constraint satisfaction requires large LP relaxations. In: Proceedings of 54th IEEE FOCS, pp. 350–359 (2013)
4. Fiorini, S., Massar, S., Pokutta, S., Tiwary, H.R., de Wolf, R.: Exponential lower bounds for polytopes in combinatorial optimization. J. ACM **16**(2) (2015). arXiv/1111.0837. (Earlier version in STOC 2012)
5. Lee, J.R., Raghavendra, P., Steurer, D.: Lower bounds on the size of semidefinite programming relaxations. In: Proceedings of 47th ACM STOC, pp. 567–576 (2015)
6. Rothvoß, T.: The matching polytope has exponential extension complexity. In: Proceedings of 46th ACM STOC, pp. 263–272 (2014)
7. Yannakakis, M.: Expressing combinatorial optimization problems by linear programs. J. Comput. Syst. Sci. **43**(3), 441–466 (1991). (Earlier version in STOC 1988)

Contents

The Shortest Separating Cycle Problem

Esther M. Arkin[1], Jie Gao[1], Adam Hesterberg[2], Joseph S.B. Mitchell[1(✉)],
and Jiemin Zeng[1]

[1] Stony Brook University, Stony Brook, NY, USA
{esther.arkin,jie.gao,joseph.mitchell,jiemin.zeng}@stonybrook.edu
[2] Massachusetts Institute of Technology, Boston, MA, USA
achester@mit.edu

Abstract. Given a set of pairs of points in the plane, the goal of the
shortest separating cycle problem is to find a simple tour of minimum
length that separates the two points of each pair to different sides. In
this article we prove hardness of the problem and provide approximation
algorithms under various settings. Assuming the Unique Games Conjecture, the problem cannot be approximated within a factor of 2. We provide a polynomial algorithm when all pairs are unit length apart with
horizontal orientation inside a square board of size $2 - \varepsilon$. We provide
constant approximation algorithms for unit length horizontal or vertical
pairs or constant length pairs on points laying on a grid. For pairs with no
restriction we have an $O(\sqrt{n})$-approximation algorithm and an $O(\log n)$-
approximation algorithm for the shortest separating planar graph.

Keywords: Shortest separating cycle · Traveling salesman problem

1 Introduction

Given a set $P = \{(p_i, q_i) | 1 \leq i \leq n\}$ of pairs of points in the plane, we seek
a *shortest separating cycle* T, a tour where for every pair of points, one point
is inside the tour and the other is outside (Fig. 1). Each pair (p_i, q_i) can be
represented by a line segment connecting p_i and q_i. Therefore each segment is
cut by the tour an odd number of times. Throughout this paper, we use whichever
interpretation is more intuitive.

The motivation for this problem originates from data storage and retrieval in
a distributed sensor network [16,18]. Consider an application in which sensors are
installed at parking spots to detect if the spot is empty, while mobile users roam-
ing around in the city are in need of such information. We would need to have a
data processing, storage and retrieval scheme to allow mobile users anywhere to
quickly retrieval data of interest. The solution of always delivering the query to
the data source may suffer from a single point of failure and traffic bottleneck.
Therefore a natural solution is to adopt geographical hashing. In [18] each data
is hashed to two storage sensors by its type. While a piece of data i is delivered
from one storage site p_i to the other storage location q_i using multi-hop routing
it is convenient for all the nodes on the relay path to also cache the data item.

© Springer International Publishing AG 2017
K. Jansen and M. Mastrolilli (Eds.): WAOA 2016, LNCS 10138, pp. 1–13, 2017.
DOI: 10.1007/978-3-319-51741-4_1

For a mobile user seeking data of a particular type i, the user can issue a query which only needs to visit a node that has cached the data. Say if the user query travels along a tour that separates p_i and q_i, the query will hit the cached data for sure – as any path connecting p_i and q_i is intersected by the tour. In this retrieval scheme one can easily query for a collection of data of multiple types $1, 2, \cdots n$, as long as the query follows a tour that separates each pair of nodes p_i and its corresponding hashed storage node q_i. This becomes precisely the separating cycle problem [18]. Finding the shortest separating cycle is natural, as the shortest tour minimizes energy consumption and delay.

In computational geometry, this problem is related to many traveling salesman problem (TSP) variants, including the red-blue separation problem, TSP with neighborhoods, and one-in-a-set TSP (also known as group TSP), that have been well studied [2,12]. All of these problems are known to be NP-hard in the Euclidean plane as they all contain the classical TSP as a special case. The shortest separating cycle problem is different from any of these problems. In

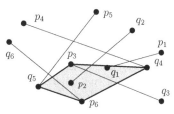

Fig. 1. The shortest separating cycle problem.

the red-blue separation problem, given a set of red and blue points in the plane, the aim is to find the shortest tour that separates the blue points from the red points. In our problem, the points in the pairs need to be separated but they are not assigned colors. Thus part of the challenge is to determine which point of each pair is inside the tour and which one is outside.

In the TSP with neighborhoods (TSPN), given a set of regions, the goal is to find the shortest tour that visits each region. One may attempt to connect a line segment for each pair in our input and apply an algorithm for TSP with neighborhoods where the neighborhoods are line segments. However, this does not necessarily give a valid solution since the TSP with neighborhoods solution might reflect on the edge and not enclose an endpoint in its cycle such as in Fig. 2(i).

For the one-in-a-set TSP, we are given a collection of sets and the problem asks for the shortest tour that visits at least one element in each set. Any one-in-a-set TSP solution to our input can be easily modified to become a separating cycle. However, a separating cycle does not need to visit every point it is including or excluding so the one-in-a-set TSP solution may be excessively long. An example of this can be seen in Fig. 2(ii).

Our Results. In this paper we are the first to study the shortest separating cycle problem and provide both hardness and approximation results. In particular, we consider special cases where the orientation of the input pairs, the distance between the input pairs, and the configuration of the domain are restricted. We vary the size of the square board the input points are confined within as well as the range of orientations the input pairs have from strictly horizontal and/or vertical to any orientation. Some cases have additional restrictions such as how far each pair of points are from each other and whether or not the input points must lie on a grid. The results are summarized in Table 1.

In general, despite the apparent similarity and connection to many other TSP variants which have easy constant-factor approximation results, the shortest separating cycle problem is a lot harder. Many ideas that were used in typical TSP algorithms are not applicable here. Indeed, Indeed, we show that the problem is hard to approximate for a factor of 1.3606 unless P = NP and is hard to approximate better than a factor of 2 assuming the Unique Games Conjecture. We provide a polynomial time algorithm when all pairs are unit length horizontal segments inside a square board of size $2 - \varepsilon$. We provide approximation algorithms for unit length horizontal or vertical segments or constant length segments on points laying on a grid. These scenarios are of particular interest to the application setting in a sensor network. Last, for arbitrary pairs we have an $O(\sqrt{n})$-approximation algorithm and an $O(\log n)$-approximation algorithm for the shortest separating planar graph problem, in which the objective is to compute an embedded planar graph of minimum total edge length so that the two endpoints of each pair are in different faces.

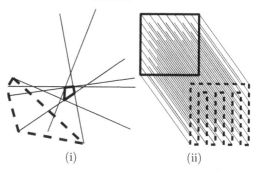

(i)   (ii)

Fig. 2. Shortest separating cycle is different from TSPN or one-in-a-set TSP. (i) The TSP with neighborhoods solution (solid) is not a valid separating cycle and is a much shorter tour than the shortest separating cycle (dashed). (ii) The one-in-a-set TSP solution (dashed) is a much longer tour than the shortest separating cycle (solid).

Table 1. Approximation algorithm and hardness results for different settings.

Board size	Unit length horizontal	Unit length horizontal & vertical	Unit length arbitrary orient	Constant length arbitrary orient. Points on grid
$2 - \varepsilon$	in P	4-approx	NP-hard	NA
$M = O(1)$	$O(1)$-approx	$(M^2 + 1)/4$-approx	Hard to approx	NA
n	$O(1)$-approx	$O(1)$-approx	Hard to approx	$O(1)$-approx

Related Work

TSP. The traveling salesman problem is one of the most well known geometric problems in history. It is one of the first problems known to be NP-hard [7,15]. In a metric setting Christofides provided a 3/2 approximation algorithm [5]. In the Euclidean setting, the problem is known to admit a PTAS, independently shown by Arora [2] and Mitchell [12].

TSP with Red Blue Separation. The red blue separation problem in the plane admits a PTAS (by [12] or by [3]).

TSP with Neighborhoods. TSP with neighborhoods was first studied by Arkin and Hassin [1] in which $O(1)$-approximation algorithms were developed when the neighborhoods are translates of a convex polygon or when the neighborhoods are unit disks. For general (nondisjoint) connected neighborhoods, an $O(\log n)$-approximation algorithm is known where n is the number of neighborhood regions [8,11], and it is NP-hard to approximate within a $2 - \varepsilon$ ratio [17]. For fat regions of bounded depth, there is a PTAS [13] (even in doubling metrics [4]), while for general connected regions of bounded depth, or for convex regions, an $O(1)$-approximation is known in two dimensions [14].

One-of-a-set TSP or Group TSP. The one-of-a-set or group TSP is the TSPN in which the neighborhoods are discrete sets of points (and thus disconnected). Safra and Schwartz [17] show the 2D problem is NP-hard to approximate to within any constant factor; for groups that are sets of k points, they also give approximation lower bounds $(\Omega(\sqrt{k}))$. Slavík [19] gives a $(3/2)k$-approximation, based on linear programming methods.

2   Hardness

The shortest separating cycle problem is NP-hard by a trivial reduction from the traveling salesman problem (TSP). For any TSP instance with cities at location w_i, we place a pair of points p_i, q_i very close to each w_i. In order to separate the points, the tour will need to visit each city w_i. Thus the shortest separating cycle problem is as hard as TSP. In the following we show stronger results that the problem is hard to approximate.

2.1   Inapproximability for Any Length Segments

Theorem 1. *The shortest separating cycle problem with no restrictions on the distance and orientation of input pairs is NP-hard to approximate better than a factor of 1.3606. It is hard to approximate better than a factor of 2 assuming the unique games conjecture.*

Proof. Our reduction is from minimum vertex cover. Given a graph $G = (V, E)$, the goal of the minimum vertex cover problem is to find a minimum cardinality subset of vertices $V^* \subseteq V$ such that every edge in E is incident to at least one vertex in V^*.

Now we will create a set of pairs in the plane. First we place each vertex in V along a circle with a center w' and designate another location w'' at a distance $\Omega(nm)$ from w'. Here, $n = |V|$ and m is a constant. For every vertex $v \in V$, we place m endpoints overlapping a single point at w' and their corresponding endpoints in a $\sqrt{m} \times \sqrt{m}$ grid pattern around v (as shown by dark dots in Fig. 3). Finally, w' and w'' are connected by a long segment. The result is a "wheel" composed of vertices and edges in G with "spokes" towards the center, a hub at w', and a large arm from w' to w''.

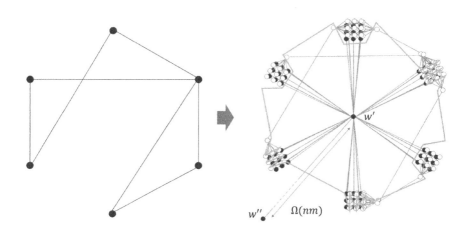

Fig. 3. Reduction from vertex cover.

For every edge $e(u, v) \in E$, we place m endpoints in a $\sqrt{m} \times \sqrt{m}$ grid pattern in a $\sqrt{m} - 1$ square around u and their corresponding endpoints overlapping a single point u_e near the grid. Another grid of m endpoints is placed in a similar grid around v with their corresponding points overlapping on a point v_e. These points are shown in hallow dots in Fig. 3. These hallow grid points are extremely close to the dark grid points. Finally, u_e and v_e are connected by a segment. Therefore, for every gadget that represents an edge, exactly one set of grid points must be inside the cycle.

Let's first consider the super long arm from w' to w'', and the pairs created by edges of G. First the separating cycle will need to include some points from the edge gadgets. If the separating cycle also visits w'' (to include it inside), there is an additional cost of length of $O(nm)$, which is so prohibitive and must be avoided. So the separating cycle will include the points at w' in the interior to separate w' and w''. This implies that the dark dots connected to w' by spokes need to be outside the cycle. Now let's consider a set of hollow grid points near a set of dark grid points. As the dark grid points need to be outside, then the hallow points need to be visited invidually to include them inside the cycle. Since points in a grid are unit distance away from each other, a path that visit a grid of points has a length of at least m.

The optimal solution visits the minimum number of vertices in V while also separating all pairs of points which is a minimum vertex cover of G. The parity of the segment chains ensure that when one point of a grid is collected, all points in that grid must be collected.

The length of an optimal separating cycle in this instance is $O(|OPT_{VC}|m + n\sqrt{m})$ where $|OPT_{VC}|$ is the size of the optimal solution of the corresponding vertex cover problem. If $m \geq n$, then the cost of navigating between grids at the vertices of V dominates the cost of the optimal solution. The rest of the cost, $O(n\sqrt{m})$ is from traveling between vertices and collecting w'.

Now we assume that we are given a δ-approximation to our construction of the shortest separator problem. This path has distance at most $O(\delta|OPT_{VC}|m + \delta n\sqrt{m})$. To convert this solution into a solution for the corresponding vertex cover problem, any grid points collected at a vertex translates to a vertex selected in the vertex cover. This means that any additional vertex above the optimal solution of the vertex cover problem translates to an additional $O(m)$ length in the given approximation of the shortest separator instance. If we let $m = \Omega(\delta^2 n^2)$, then the approximation solution visits at most $\delta|OPT_{VC}|$ vertices and is a δ-approximation for vertex cover. Therefore, the lower bounds for the vertex cover problem apply to the shortest separator problem. This problem cannot be approximated with a factor better than 1.3606 unless $P = NP$ or 2 assuming the unique games conjecture.

3   Algorithms

We describe exact and approximation algorithms for the shortest separating cycle problem under different scenarios.

3.1   Board Size $2 - \varepsilon$, Horizontal and Vertical Unit Segments

In this scenario, all n input segments are inside a square board of size $2 - \varepsilon$ for some $\varepsilon > 0$ and are restricted to have horizontal or vertical orientations. Without loss of generality we assume all the endpoints of different input segments do not share a common x-coordinate or y-coordinate. This can be done by perturbing the input slightly.

First, all unit length boxes which contain *at least* one endpoint of each segment are found. This can be executed in polynomial time by checking all possible combinatorial configurations of unit length boxes. The total number of combinatorial types of such squares is $O(n^2)$ since we can assume without loss of generality that the square always has two input endpoints (from two different segments) on its boundary. Aside from the two points on the boundary, each such box actually contains *exactly* one endpoint of each input segment. For the two boudary points we enumerate all combinations of including these boundary points inside the box. The convex hull of all endpoints inside the box is a candidate separating cycle. We can enumerate all such boxes to find the shortest separating cycle.

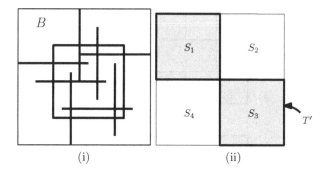

Fig. 4. The two cases in our algorithm for the shortest separating cycle on a $2 - \varepsilon$ board. Case (i): Exactly one endpoint of each input pair can be enclosed in a unit square. Case (ii): The curve T' traverses around S_1 and S_3.

If a unit length box that strictly contains one endpoint of every segment cannot be found, then the board is divided into four squares each of size $1 - \varepsilon/2$ and are colored in a checker board pattern. The square are named S_1, S_2, S_3, S_4 in a counter clockwise manner. Consider two squares along the diagonal (named S_1 and S_3, see Fig. 4(ii)). We create a tour that walks along the perimeter of their union. To accommodate corner cases, we consider the top and left border of S_3 to be open edges. This generates a curve T' of length $8 - 4\varepsilon$.

Theorem 2. *For the shortest separating cycle problem with a square $2 - \varepsilon$ domain where input pairs are exactly one apart and have either horizontal or vertical orientation, our algorithm outputs a cycle that is a 4-approximation to the optimal solution.*

Proof. There are two cases in the algorithm which outputs two different types of cycles. In case (i), the optimal solution fits inside a unit length box B. Every segment has one endpoint inside B and B will be discovered in the first phase of the algorithm. The convex hull of the points inside B is the shortest separating tour that contains all points in B.

For case (ii) we know the optimal tour cannot be completely contained by a unit box and therefore must have length at least 2. We first argue that T' is a valid separating cycle. Since each segment has unit length, the two endpoints of each segment cannot fit inside any single square of size $1 - \varepsilon/2$ and thus cannot both lie inside $S_1 \cup S_3$ nor inside $S_2 \cup S_4$. Therefore each segment must have exactly one endpoint inside $S_1 \cup S_3$. Therefore, T' is a valid separating cycle. Since T' has length $8 - 4\varepsilon$ and the optimal tour has length at least 2, T' is a 4-approximation of the optimal solution.

3.2 Constant-Size Boards

We can extend the checkerboard strategy for any constant board size M, where M is an integer. The only modification is that in the second case, a larger

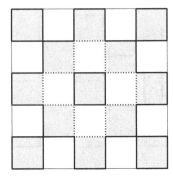

Fig. 5. Constant sized square board with cycles along the perimeter of the dark squares. (Color figure online)

checkerboard of $M \times M$ unit squares is used. The squares are colored in a checkerboard manner, and partitioned into white squares and dark squares. To make the tiling a perfect partition of the plane, we consider each square pixel to include its top edge, except for the NE corner, and to include its left edge, except for the SW corner. Again any unit length vertical or horizontal segment has two endpoints in different colored squares. Thus a tour T' that separates the white squares from the dark squares would be a valid separating cycle. Such a tour can always be found by taking a cycle along the boundary of the outermost ring of the dark squares, and iterating towards the center. All the tours can be joined into a single tour of the same length. See Fig. 5 for an illustration.

Theorem 3. *We consider the case of the shortest separating cycle problem with an $M \times M$ square domain and where the input pairs are restricted to be exactly one apart with either horizontal or vertical orientation. Our algorithm including the checkerboard strategy is an $(M^2 + 1)/4$-approximation.*

Proof. Our proof is similar to the proof of Theorem 2. Either we can find a unit length box containing at least one endpoint of each segment (in which case we find the optimal solution), or the optimal tour has length at least 2. In the second case we will take the tour T' along the perimeter of the union of the dark squares. T' has length at most $M^2/2$ if M is even, and at most $(M^2 + 1)/2$ if M is odd. Thus the approximation factor is at most $(M^2 + 1)/4$.

3.3 Any Board Size, Horizontal and Vertical Unit Segments

If the board size M is a constant, we can apply the same checkboard idea as in the previous section. But when M is large, we have to use a different idea to get a constant approximation.

First, we overlay a grid of unit squares over the domain partitioning the domain into light and dark squares in a checkerboard pattern. For each grid cell, we consider the top edge, excluding the NE corner, and the left edge, excluding the SW corner, as closed edges.

We refer to dark squares that have a point (from a pair) in them as "occupied". Let S be the occupied squares. Let S' be the 3-by-3 squares centered on the squares of S. In the following we assume without loss of generality that $|S| \geq 5$. The case $|S| \leq 4$ can have an arbitrarily small optimal value; but this constant-size case can be easily handled.

Now we consider the shortest TSP with Neighborhoods (TSPN) tour on the set S' of enlarged squares and name the length as $TSPN(S')$. This tour connects the regions in S' but we must also separate the pairs of points in each region. We further apply the constant factor approximation to each region in S'. Our algorithm is simply this: Run a TSPN algorithm on S' (for which there is a PTAS [6]), and augment the tour with our approximation algorithm for square regions of constant size.

Theorem 4. *Our algorithm for the shortest separating cycle problem for input pairs restricted to a separation of exactly 1 and only horizontal or vertical orientation, is a constant approximation.*

Proof. We have two cases regarding the size of S, $|S| \geq 5$ and $|S| \leq 4$. In the first case, $|S| \geq 5$, the length of the output is at most $(1 + \varepsilon)TSPN(S') + O(|S|)$. Since the optimal solution must visit every enlarged square in S', then $OPT = \Omega(TSPN(S'))$. Assuming no single point stabs all squares of S' (i.e., assuming $|S| \geq 5$), the standard packing argument shows that TSPN on a set of nearly disjoint (i.e., constant depth of overlap), equal-sized squares requires length proportional to the number of squares times the side length of the squares. This leads us to claim that $OPT = \Omega(|S|)$. Therefore, $(1+\varepsilon)TSPN(S')+O(|S|)$ is $O(OPT)$ and our tour is a constant approximation.

For the case where $|S| \leq 4$, our strategy only changes if a single point is contained in all squares of S'. In this case, our entire input can be contained in a 5×5 square and we refer to our algorithm for constant sized domains.

3.4   Any Orientation, Bounded Aspect Ratio

We now assume that the distance between any two points (not restricted to designated pairs) in S are greater than or equal to 1. The segments defined by the pairs of points may have any orientation and their distances are bounded by a constant. Let $r = cL$ for some constant value of $c \geq 1$ where L is the length of the longest segment. The aim is to find a subset I of pairs of points where the shortest distance between any two segments is greater than r and the shortest distance between any input segment and it's closest segment in I is less than r. Then we find the TSPN path on line segments in I. Next, we divide the region into neighborhoods by assigning the remaining segments to their nearest segment in I. The TSPN tour of the segments, $TSPN(I)$, is augmented with detours that separate all of the segments in each neighborhood. The resulting tour is a constant approximation of the optimal separating tour.

To find such an independent set, we randomly select neighborhoods of size $O(r)$ until all segments have been selected. A segment s is randomly chosen and removed from the set of input segments along with all remaining segments within distance (shortest distance) r of s. The segment s is placed in the set I. This procedure is repeated until all segments are removed. The shortest distance between any two segments in I is greater than r. The shortest distance between any segment and it's closest segment in I is less than or equal to r. Each segment is assigned to it's closest segment in I. The set of segments assigned to a segment s in I is denoted as $N(s)$.

A tour is constructed by first finding a TSPN tour on I. Then the path is augmented by a shorter tour within each neighborhood of each segment in I. When a tour reaches a segment s in I, then it makes a separating detour that separates all of the segments in $N(s)$. The length of a tour is bounded by $O(L^2)$ since all of the segments in $N(s)$ is within a neighborhood of radius $2r$ of s and by a packing argument, there are $O(L^2)$ possible points in such a neighborhood. In the worst case, the separating tour visits every point in the neighborhood and includes or excludes each point as required. Note that the detour must exclude segments that are not in $N(s)$.

Theorem 5. *The path our algorithm produces is a $O(L^2)$ approximation of the optimal minimum perimeter separator.*

Proof. Let T be the length of the path our algorithm produces, let OPT be the length of the optimal solution and let $TSPN(I)$ be the length of the TSPN path on I. Our path is a separating tour because every segment is separated by the tour within its neighborhood and excluded by the tour everywhere else. We claim the length of such a tour is bounded above by $TSPN(I) + O(|I|L^2)$. Since the TSPN path of I must enter and exit the neighborhood of every segment in I, $TSPN(I) = \Omega(|I|)$. Therefore $T = O(L^2 \cdot TSPN(I))$. Since $OPT = \Omega(TSPN(I))$ and $T \geq OPT$, then $T = O(L^2 \cdot OPT)$.

3.5   The General Case

For general pairs of points in the plane, we observe that an $O(\sqrt{n})$-approximation follows from known results on the Euclidean TSP in the plane. Specifically, we first compute a minimum-size square, Q, that contains at least one point of each pair. (This is easily computed, since the n point pairs determine only $O(n^3)$ combinatorially distinct squares.) Now, within Q, we compute an approximate TSP tour T (using any constant-factor approximation method for TSP) on the points that are inside (or on the boundary of) Q, making sure the approximate tour is a simple polygon. We obtain a valid separating cycle for the input pairs as follows: Consider traversing T, starting from an arbitrary point. Each time we reach a point along this traversal, we either make a slight detour to include it (if it is the first time we have encountered a point from this pair), or make a slight detour to exclude it (if it is the second encounter with a pair). In this way, we obtain a valid separating cycle just slightly longer than T. By classic results on the Euclidean TSP (see, e.g., Karloff [9]), we know that the length of T is at most $O(|Q|\sqrt{n})$, where $|Q|$ is the side length of the square Q. Since we know that $\Omega(|Q|)$ is a lower bound on the length of an optimal separating cycle, we have shown that in polynomial time one can obtain an $O(\sqrt{n})$-approximation for the general case of our problem.

3.6   Separating Subdivision Problem

We consider now a different version of the separating cycle problem – the *separating subdivision problem*, in which the goal is to compute an embedded planar graph of minimum length such that every input pair has its points in different faces of the subdivision. We define the length of the graph to be the sum of the lengths of all of its edges. We give an $O(\log n)$-approximation algorithm. We outline the approach, deferring details to the full paper. We argue that an optimal subdivision, S, can be converted to a special (recursive) "guillotine" structure, increasing its length by a factor $O(\log n)$; then, we show that an optimal solution among guillotine structures can be computed in polynomial time, using dynamic programming. The conversion goes as follows. First, increasing the total edge length of S by at most a constant factor, we can convert its faces to all be rectilinear: we enclose S with its bounding box, and replace each face of S with a rectilinear polygon, with axis-parallel edges that lie on the grid induced by the input point pairs, while keeping all points within their respective faces. Then, we partition each simple rectilinear face into rectangles, adding axis-parallel chords that lie on the grid; this causes the total edge length to go up by a factor $O(\log n)$; see [10]. Then, using the charging scheme of [10], we know that we can convert the resulting *rectangular subdivision* to a *guillotine rectangular subdivision*, in which one can recursively partition the subdivision using axis-parallel "guillotine" cuts that do not enter the interior of rectangular faces. Optimizing the length of a guillotine rectangular subdivision is done with dynamic programming, in which subproblems are axis-aligned rectangles all of whose boundary is (by definition) included in the edge set of the subdivision. This implies that any input pair of points that "straddles" the boundary of a subproblem, with one point inside, one point outside, is already satisfied automatically with respect to pair separation (the points lie in different faces/rectangles). This means that the subproblem is only responsible for the separation of the point pairs both of whose points lie within the defining rectangle of the subproblem. The algorithm computes a minimum-length guillotine rectangular subdivision, separating all point pairs. Since an optimal solution can be converted to the class of guillotine rectangular subdivisions at a lengthening factor $O(\log n)$, we obtain the claimed approximation.

4   Conclusion and Future Work

The shortest separating cycle is a new variant of the TSP family that has not been studied before. This paper provides the first set of hardness bounds and a number of approximation algorithms under different settings. The gap for the approximation ratios and hardness results is still big and narrowing or closing the gap is the obvious future work.

Acknowledgement. E. Arkin and J. Mitchell acknowledge support from NSF (CCF-1526406). J. Gao and J. Zeng acknowledge support from AFOSR (FA9550-14-1-0193) and NSF (CNS-1217823, DMS-1418255, CCF-1535900).

References

1. Arkin, E.M., Hassin, R.: Approximation algorithms for the geometric covering salesman problem. Discrete Appl. Math. **55**(3), 197–218 (1994)
2. Arora, S.: Polynomial time approximation schemes for Euclidean traveling salesman and other geometric problems. J. ACM **45**(5), 753–782 (1998)
3. Arora, S., Chang, K.: Approximation schemes for degree-restricted MST and red-blue separation problems. Algorithmica **40**(3), 189–210 (2004)
4. Chan, T.H., Jiang, S.H.: Reducing curse of dimensionality: improved PTAS for TSP (with neighborhoods) in doubling metrics. In: Proceedings of the Twenty-Seventh Annual ACM-SIAM Symposium on Discrete Algorithms, SODA 2016, 10–12 January 2016, Arlington, VA, USA, pp. 754–765 (2016)
5. Christofides, N.: Worst-case analysis of a new heuristic for the travelling salesman problem. Technical report 388, Graduate School of Industrial Administration, Carnegie Mellon University (1976)
6. Dumitrescu, A., Mitchell, J.S.: Approximation algorithms for TSP with neighborhoods in the plane. J. Algorithms **48**(1), 135–159 (2003). Twelfth Annual ACM-SIAM Symposium on Discrete Algorithms
7. Garey, M.R., Graham, R.L., Johnson, D.S.: Some NP-complete geometric problems. In: Proceedings of the Eighth Annual ACM Symposium on Theory of Computing, STOC 1976, pp. 10–22. ACM, New York (1976)
8. Gudmundsson, J., Levcopoulos, C.: Hardness result for TSP with neighborhoods. Technical report, Technical Report LU-CS-TR: 2000–216, Department of Computer Science, Lund University, Sweden (2000)
9. Karloff, H.J.: How long can a Euclidean traveling salesman tour be? SIAM J. Discrete Math. **2**(1), 91–99 (1989)
10. Mata, C., Mitchell, J.S.B.: Approximation algorithms for geometric tour and network design problems. In: Proceedings of the 11th Annual ACM Symposium on Computational Geometry, pp. 360–369 (1995)
11. Mata, C.S., Mitchell, J.S.B.: Approximation algorithms for geometric tour and network design problems (extended abstract). In: Proceedings of the Eleventh Annual Symposium on Computational Geometry, SCG 1995, pp. 360–369. ACM, New York (1995)
12. Mitchell, J.S.B.: Guillotine subdivisions approximate polygonal subdivisions: a simple polynomial-time approximation scheme for geometric TSP, k-MST, and related problems. SIAM J. Comput. **28**(4), 1298–1309 (1999)
13. Mitchell, J.S.B.: A PTAS for TSP with neighborhoods among fat regions in the plane. In: Proceedings of the Eighteenth Annual ACM-SIAM Symposium on Discrete Algorithms, SODA 2007, pp. 11–18 (2007)
14. Mitchell, J.S.B.: A constant-factor approximation algorithm for TSP with pairwise-disjoint connected neighborhoods in the plane. In: Proceedings of the 26th Annual ACM Symposium on Computational Geometry, pp. 183–191 (2010)
15. Papadimitriou, C.H.: The Euclidean travelling salesman problem is NP-complete. Theor. Comput. Sci. **4**(3), 237–244 (1977)

16. Ratnasamy, S., Karp, B., Yin, L., Yu, F., Estrin, D., Govindan, R., Shenker, S.: GHT: a geographic hash table for data-centric storage in sensornets. In: Proceedings 1st ACM Workshop on Wireless Sensor Networks ands Applications, pp. 78–87 (2002)
17. Safra, S., Schwartz, O.: On the complexity of approximating TSP with neighborhoods and related problems. Comput. Complex. **14**(4), 281–307 (2006)
18. Sarkar, R., Zhu, X., Gao, J.: Double rulings for information brokerage in sensor networks. IEEE/ACM Trans. Netw. **17**(6), 1902–1915 (2009)
19. Slavík, P.: The errand scheduling problem. Technical report 97-2, Department of Computer Science, SUNY, Buffalo (1997)

Dynamic Traveling Repair Problem with an Arbitrary Time Window

Yossi Azar$^{(\boxtimes)}$ and Adi Vardi$^{(\boxtimes)}$

School of Computer Science, Tel-Aviv University, 69978 Tel-Aviv, Israel
azar@tau.ac.il, adi.vardi@gmail.com

Abstract. We consider the online Dynamic Traveling Repair Problem (DTRP) with an arbitrary size time window. In this problem we receive a sequence of requests for service at nodes in a metric space and a time window for each request. The goal is to maximize the number of requests served during their time window. The time to traverse between two points is equal to the distance. Serving a request requires unit time. Irani et al., SODA 2002 considered the special case of a fixed size time window. In contrast, we consider the general case of an arbitrary size time window. We characterize the competitive ratio for each metric space separately. The competitive ratio depends on the relation between the minimum laxity (the minimum length of a time window) and the diameter of the metric space. Specifically, there exists a constant competitive algorithm only when the laxity is larger than the diameter. In addition, we characterize the rate of convergence of the competitive ratio, which approaches 1, as the laxity increases. Specifically, we provide matching lower and upper bounds. These bounds depend on the ratio between the laxity and the optimal TSP solution of the metric space (the minimum distance to traverse all nodes). An application of our result improves the previously known lower bound for colored packets with transition costs and matches the known upper bound. In proving our lower bounds we use an embedding with some special properties.

1 Introduction

Consider an employee in the Google IT division. He is responsible for replacing malfunctioning disks in Google's huge computer farms. During his shift he receives requests to replace disks at some points in time. Each request is associated with a deadline. If the disk will not be replaced before the deadline, there is a high probability that the performance of the Search Engine will experience a significant hit. Replacing a disk takes unit time (service time). However, before the employee can replace it, he must travel from his current location to the location of the disk. The goal is to maximize the number of disks replaced before their deadline. What path should the employee take and how should the path change with new requests? Irani et al., SODA 2002 [15,18] called this online

---

Supported in part by the Israel Science Foundation (grant No. 1506/16), by the Israeli Centers of Research Excellence (I-CORE) program, (Center No. 4/11) and by the Blavatnik Fund.

K. Jansen and M. Mastrolilli (Eds.): WAOA 2016, LNCS 10138, pp. 14–26, 2017.
DOI: 10.1007/978-3-319-51741-4_2

problem the Dynamic Traveling Repair Problem (DTRP). They considered the special case of a fixed size time window, where the window of a request is the period between its release time and its deadline. In contrast, we consider the general case of an arbitrary size time window. In this paper we characterize the competitive ratio for each metric space separately. We determine whether the competitive ratio is constant or not depending on the minimum laxity (the minimum length of a time window) and the diameter of the metric space (the maximum distance between nodes in the metric space). In addition, we consider the case where the laxity is large compared to the optimal TSP solution of the metric space (the minimum distance to traverse all nodes). Specifically, we provide matching lower and upper bounds for these cases. These bounds depend on the ratio between the laxity and the optimal TSP solution of the metric space.

We note that even when the service time is not negligible, our problem can be reduced to TSP with time windows and zero service time [5] by changing the metric space. However, our competitive ratio depends on the properties of the metric space and the reduction might change the parameters of the metric space significantly. Hence, it might influence a crucial parameter which determines the competitive ratio. Therefore, we take service time into account in our model. Moreover, in our main result, where the laxity is larger than the optimal TSP solution of the metric space, without service time it is easy to design a 1-competitive algorithm by traveling over an optimal TSP solution periodically.

Offline Problem. Note that in the offline case (i.e., when the sequence is known in advance), if the service time is negligible compared to the minimum positive distance between nodes (or 0) then the problem becomes TSP (or vehicle routing) with time windows and zero service time [5]. Moreover, if in addition all deadlines are the same and all release times are zero then the problem reduces to the (offline) orienteering problem [1, 3, 14]. Vehicle routing problem (with time windows and zero service time) has been extensively studied both in computer science and the operations research literature, see [11, 12, 19–22]. For an arbitrary metric space Bansal et al. [5] showed an $O(\log^2 n)$-approximation (for certain cases a better approximation can be achieved [8]). Constant factor approximations have been presented for the case of points on a line [6, 17, 23]. For the orienteering problem, i.e., all release times are zero, all deadlines are the same, and the service time is zero, there are constant factor approximation algorithms [5, 7, 9, 10]. A restricted online version of the Vehicle Routing problem (without deadlines) was considered in [2, 13, 16].

Application for Packet Scheduling. Another motivation for our problem is the Colored Packets with Deadlines and Metric Space Transition Cost problem. In this setting we are given a sequence of incoming colored packets. Each colored packet is of unit size and has a deadline. There is a reconfiguration cost (setup cost) to switch between colors (the cost depends on the colors). The goal is to find a schedule that maximizes the number of packets that are transmitted before the deadline. Note that for one color the earliest deadline first (EDF) strategy is known to achieve an optimal throughput. The unit cost color has been considered in [4]. In particular, when we apply our results to the uniform metric space we improve the previous lower bound and match the known upper bound.

1.1   Our Results

Denote by σ the sequence of requests. The window of request i is $[r_i, d_i]$, where r_i is the release time of the request and d_i is the deadline of the request. Let $L = \min_{i \in \sigma}\{d_i - r_i\} \geq 1$ be the minimum laxity of the requests (the minimum length of a time window). Note that the laxity has to be at least 1 since the service time equals 1. Denote by $\Delta(G)$ the diameter of the metric space G, i.e., the largest distance between two nodes. Denote by $TSP(G)$ the weight of a minimal TSP solution in the metric space G and $MST(G)$ the weight of a minimal spanning tree.

In this paper we characterize when it is possible to achieve a $\Theta(1)$ competitive algorithm for the Dynamic Traveling Repair Problem with an arbitrary time window, and when the best competitive algorithm is unbounded. Moreover, we characterize the rate of convergence of the competitive ratio, which approaches 1 as the laxity increases. Specifically, we provide matching lower and upper bounds depending on the ratio between the laxity and the optimal TSP solution of the metric space.

It is also interesting to mention that in many cases the competitive ratio of an algorithm is computed as the supremum over all metric spaces while lower bounds are proved for one specific metric space. In contrast, we prove more refined results. Specifically, we show an upper bound and a lower bound for each metric space separately. Hence, one cannot design a better competitive algorithm for the specific metric space that one encounters in the real specific instance. Hence, even for specific metric spaces, we show it is impossible to do better.

We consider three cases. The last two cases are done for completeness of the result while the first case is our main result.

- **Case A:** $L > TSP(G)$. Let $\delta = TSP(G)/L < 1$. We show a strictly larger than 1 lower bound. Specifically, if $\delta \leq \frac{1}{256}$ we provide a lower bound of $1 + \Omega(\sqrt{\delta})$ as well as a matching upper bound of $1 + O(\sqrt{\delta})$.
  We note that without service time it is easy to design 1-competitive algorithm by traveling over an optimal TSP solution periodically. Recall that there is a reduction from the service time model to a model without service time that seems to contradict the lower bound (see [5]). However, the reduction modifies the metric space and hence increases δ such that δ is not smaller than $\frac{1}{256}$.
- **Case B:** $3\Delta(G) < L \leq TSP(G)$. We design a $O(1)$-competitive algorithm and a 1.00054 lower bound.
- **Case C:** $L < \Delta(G)/2$. For any metric space the competitive ratio of any deterministic online algorithm is unbounded (easily proved). For randomized algorithms the competitive ratio depends on the metric space. For example, for a metric space which consists of 2 points one can easily show a 4-competitive algorithm even for $L = 0$. In contrast, in a uniform metric space the competitive ratio is at least $|V|$ where V is the number of nodes in the metric space, even for $L < \Delta(G)$.
  Note that in the remaining cases, i.e., $\Delta(G)/2 \leq L \leq 3\Delta(G)$, the question of whether there exists a constant competitive algorithm depends on the metric space for both deterministic and randomized algorithms. Specifically, for

deterministic algorithms where $L = \Delta(G)$ it is easy to prove that there is no constant competitive algorithm for the uniform metric space. In contrast, there is a constant competitive algorithm for the line metric space. As mentioned above, for randomized algorithms the bound depends on the number of nodes in the metric space for a given diameter.

Application. For the uniform metric space (when all distances are unit size), our problem is equivalent to the Colored Packets with Deadlines problem. In this case our result improves the lower bound of [4]. Specifically, we improve their $1 + \Omega(\delta)$ lower bound to $1 + \Omega(\sqrt{\delta})$ and match their upper bound for the uniform metric space.

Embedding Result. One of the techniques that we use for the lower bound is the following embedding. Let $w(S)$ denote the weight of the star metric S (i.e., the sum of the weights of the edges of S). We prove that for any given metric space G on nodes V and for any vertex $v_0 \in V$ there exists a star metric S with leaves V and an embedding $f : G \to S$ from G to S (f depends on v_0) such that:

1. $w(S) = \text{MST}(G)$.
2. The weight of every Steiner tree in S that contains v_0 is not larger than the weight of the Steiner tree on the same nodes in G.

Note that this embedding is different from the usual embedding since we do not refer specifically to distances between vertices. Typically, an embedding is used to prove an upper bound by simplifying the metric space. In contrast, our embedding is used to prove a lower bound.

In order to prove the lower bound we first establish it for the star metric, and then extend it to general metric spaces. Note that a lower bound for a sub-graph is not a lower bound for the original graph. For example, a lower bound for an MST of a metric space G is not a lower bound for G since the algorithm may use additional edges to reduce the transition time.

2   The Model

We formally model the Dynamic Traveling Repair Problem with an arbitrary time window as follows. Let $G = (V, w)$ be a given metric space where V is a set of n nodes and w is a distance function. Let $s \in V$ be a given initial node. We are given an online sequence of requests for service. Each request is characterized by a pair $([r_i, d_i], v_i)$, where $r_i \in N_+$ and $d_i \in N_+$ are the respective arrival time and deadline of the request, and $v_i \in V$ is a node in the metric space G. The time to traverse from node v_i to node v_j is $w(v_i, v_j)$. For simplicity we assume that $w(v_i, v_j)$ is integral. Serving a request at some node requires unit size service time. The goal is to serve as many requests as possible within their time windows $[r_i, d_i]$, starting from node s.

Note that when all r_i are equal to 0 and all d_i are equal to B and the service time is negligible the problem reduces to the well-known orienteering problem with budget B and a prize for each node which is equal to the number of requests at this node. That is, finding a path of total distance at most B that maximizes the prize of all visited nodes.

Let $\mathrm{ALG}(\sigma)$, $\mathrm{OPT}(\sigma)$ denote the respective throughput of the online, optimal offline algorithms with respect to a sequence σ. We consider a maximization problem and hence $\inf_\sigma \mathrm{OPT}(\sigma)/\mathrm{ALG}(\sigma) \geq 1$.

3   Lower Bounds

3.1   Lower Bound for a Small Diameter Laxity Ratio (Case A and B)

In this section we consider Cases A and B. Let $\delta = TSP(G)/L$. If $\delta < 1$ (Case A), we show a strictly larger than 1 lower bound. Specifically, if $\delta \leq \frac{1}{256}$ we provide a lower bound of $1 + \Omega(\sqrt{\delta})$. If $\delta > 1$ (Case B) we can use requests with a laxity of $256TSP(G)$ (i.e., $\delta = \frac{1}{256}$), and obtain a lower bound of 1.00054. Therefore, from now on we only consider Case A.

Lower Bound for a Star Metric. In this section we consider the case where the traveling time between nodes is represented by a star metric. This is also equivalent to the case where the traveling time from node i is w_i.

The general idea is that the adversary creates many requests with a large deadline at node v_0 at each time unit, and also blocks of fewer requests with close deadlines at other nodes. Any online algorithm must choose between serving many requests with a large deadline or traveling between many nodes and serving requests with close deadlines.

Recall that $w(S)$ denotes the weight of the star metric S (i.e., the sum of the weights of the edges of S). Let w_i denote the weight of the edge incident to vertex v_i. We define $F = \sqrt{w(S)L}$. Let $\delta = \frac{TSP(G)}{L} = \frac{2w(S)}{L}$.

Theorem 1. *No deterministic or randomized online algorithm can achieve a competitive ratio better than $1 + \Omega(\sqrt{\delta})$ for any given star metric S when $\delta \leq \frac{1}{256}$. Otherwise, if $\delta > \frac{1}{256}$, the bound becomes 1.00054.*

Proof. Let S be a given star metric with nodes $V = \{v_0, \ldots, v_{n-1}\}$. We will construct a sequence $\sigma(S, \mathrm{ALG})$ such that:

$$\frac{\mathrm{OPT}(\sigma)}{E(\mathrm{ALG}(\sigma))} \geq min \left\{ \frac{3-\delta}{3 - \frac{1}{8}\left(\sqrt{\delta/2}\right)}, \frac{3}{3 - \frac{1}{4}\left(\sqrt{\delta/2}\right)}, \frac{3}{3 - \frac{1}{48}\left(\sqrt{\delta/2}\right)} \right\}$$

Note that we can assume, without loss of generality, that $\delta \leq \frac{1}{256}$, since otherwise one may use requests with a laxity of $256w(S)$ (i.e., $\delta = \frac{1}{256}$), and obtain a lower bound of 1.00054. Let $v_0 \in V$ be a type A node and the rest of the nodes type B. Let type A requests and type B requests refer to requests at a type A node and type B node, respectively. We begin by describing the sequence $\sigma(S, \mathrm{ALG})$.

Sequence Structure: Recall that each request is characterized by a pair $([r_i, d_i], v_i)$, where $r_i \in N_+$ and $d_i \in N_+$ are the respective arrival time and deadline of the request, and v_i is a node in S. There are up to $N = \frac{L}{3F} = \frac{1}{3}\sqrt{\frac{L}{w(S)}}$

blocks, where each block consists of $3F$ time units. Let $t_i = 1 + 3(i-1)F$ denote the beginning time of block i. For each block i, where $1 \leq i \leq N$, F requests located at various nodes arrive at the beginning of the block. Specifically, $\frac{w_j}{w(S) - w_0} F$ type B requests ($[t_i, L + t_i], v_j$), for each $1 \leq j \leq n-1$, are released. A type A request ($[t, 3L], v_0$) is released at each time unit t in each block. Once the adversary stops the blocks, additional requests arrive (we call this the final event). The exact sequence is defined as follows:

1. $i \leftarrow 1$.
2. Add block i.
3. If with probability at least $1/4$ there are at least $F/2$ unserved type B requests at the end of block i (denoted by Condition 1), then L requests ($[t_{i+1}, L + t_{i+1}], v_1$) are released and the sequence is terminated. Clearly, t_{i+1} is the time of the final event. Denote this by Termination Case 1.
4. Else, if with probability at least $1/4$, at most $2F$ requests are served during block i (denoted by Condition 2), then $3L$ requests ($[t_{i+1}, 3L], v_0$) are released and the sequence is terminated. Clearly, t_{i+1} is the time of the final event. Denote this by Termination Case 2.
5. Else, if $i = N$ (there are N blocks, none of which satisfy Conditions 1 or 2) then $3L$ requests ($[L+1, 3L], v_0$) are released, and the sequence is terminated. Clearly, $L + 1$ is the time of the final event. Denote this by Termination Case 3.
6. Else ($i < N$) then $i \leftarrow i + 1$, Goto 2.

We make the following **observations:** (i) Each block consists of $3F$ time units. Hence, if ALG served at most $2F$ requests during a block, there must have been at least F idle time units. (ii) There are up to $\frac{1}{3}\sqrt{\frac{L}{w(S)}}$ blocks and each block consists of $3\sqrt{w(S)L}$ time units. Hence, the time of the final event is at most $L + 1$. (iii) Exactly one type A request arrives at each time-slot until the final event. Hence, at most L type A requests arrive before (not including) the final event. (iv) During each block, exactly F type B requests arrive, which sum up to at most $L/3$ type B requests before (not including) the final event.

Now we can analyze the competitive ratio of $\sigma(S, \text{ALG})$. Consider the following possible sequences (according to the termination type):

1. Termination Case 1: Let Y denote the number of requests in the sequence. According to the observations, the sequence consists of at most L type A requests, and at most $\frac{4}{3}L$ type B requests ($L/3$ until the final event and L at the final event). Hence, $Y \leq L + \frac{4}{3}L \leq 3L$.
   - **We bound the performance of ALG:** At time t_{i+1} there is a probability of at least $1/4$ that ALG has $L + F/2$ unserved type B requests. Since type B requests have a laxity of L, ALG can serve at most $L + 1$ of them, and must drop at least $F/2 - 1$. The expected number of served requests is

$$E(\text{ALG}(\sigma)) \leq Y - \frac{1}{4}(F/2 - 1) = Y - \frac{1}{8}F + 1/4.$$

   - **We bound the performance of an algorithm OPT′:** OPT′ serves the requests in three stages:

- **Type B requests that arrive before the final event:** Recall that all type B requests in a block arrive at once in the beginning of the block. In each block OPT$'$ first serves all requests at node v_1, then all requests at node v_2, and so on. It is clear that OPT$'$ needs at most $F + 2w(\mathrm{S})$ time units to serve the requests (F for serving and $2w(\mathrm{S})$ for traveling). OPT$'$ serves the requests starting from the beginning of the block. Recall that $L \geq 256w(\mathrm{S})$ and $F = \sqrt{w(\mathrm{S})L}$. Therefore $2F \geq 512w(\mathrm{S})$. Since the block's size is $3F$, there are enough time units. Moreover, since $L \geq 256w(\mathrm{S})$, $L \geq 16\sqrt{w(\mathrm{S})L} = 16F > F + 2w(\mathrm{S})$. Hence, all requests can be served before their deadline.
- **Type B requests that arrive during the final event:** The L requests $([t_{i+1}, L + t_{i+1}], v_1)$ that arrive during the final release time are served by OPT$'$ consecutively from time t_{i+1}. OPT$'$ can serve L requests, except for one travel phase, and hence may lose at most $2w(\mathrm{S})$ requests. According to our observations, the time of the final event t_{i+1} is at most $L + 1$. Hence, OPT$'$ serves all type B requests until time unit $2L$.
- **Type A requests:** OPT$'$ serves the L type A requests consecutively from time unit $2L+1$. Since the deadlines are $3L$, OPT$'$ serves all type A requests.

We conclude that $\mathrm{OPT}(\sigma) \geq \mathrm{OPT}'(\sigma) \geq Y - 2w(\mathrm{S})$.

The competitive ratio is

$$\frac{\mathrm{OPT}(\sigma)}{E(\mathrm{ALG}(\sigma))} \geq \frac{Y - 2w(\mathrm{S})}{Y - \frac{1}{8}F + 1/4} \geq \frac{3L - 2w(\mathrm{S})}{3L - \frac{1}{8}F + 1/4} \geq \frac{3L - 2w(\mathrm{S})}{3L - \frac{1}{8}\left(\sqrt{w(\mathrm{S})L}\right) + 1/4} = 1 + \Omega\left(\sqrt{\delta}\right).$$

Here the second inequality holds since $Y \leq 3L$, the number is above 1 and the numerator and the denominator increase by the same value.

2. Termination Case 2: The sequence consists of more than $3L$ type A requests, and all deadlines are at most $3L$.
   - **We bound the performance of ALG:** The probability that ALG was idle for F time units is at least $1/4$. Hence, the expected number of served requests is $E(\mathrm{ALG}(\sigma)) \leq 3L - \frac{1}{4}F$.
   - **We bound the performance of OPT$'$:** At each time unit until the final event, OPT$'$ serves the type A request that arrived at that particular time unit. Consequently, from the final event until time unit $3L$, OPT$'$ serves the type A requests that arrived at the final event. Therefore, OPT$'$ serves $3L$ type A requests, and so $\mathrm{OPT}(\sigma) \geq \mathrm{OPT}'(\sigma) \geq 3L$.

The competitive ratio is

$$\frac{\mathrm{OPT}(\sigma)}{E(\mathrm{ALG}(\sigma))} \geq \frac{3L}{3L - \frac{1}{4}F} = \frac{3L}{3L - \frac{1}{4}\left(\sqrt{w(\mathrm{S})L}\right)} = 1 + \Omega\left(\sqrt{\delta}\right).$$

3. Termination Case 3: the sequence consists of $3L$ type A requests, and all deadlines are at most $3L$.
   - **We bound the performance of ALG:** Let U_i be the event that the number of unserved type B requests at the end of block i is less than

$F/2$. If U_i occurs, then let j_k, $1 \le k \le r$, be the type B nodes visited by ALG in block i. At least $F/2$ requests that arrived in this block have to be served (recall that F type B requests arrive at the beginning of each block). Therefore,

$$\frac{w_{j_1}}{w(\mathrm{S}) - w_0}F + \frac{w_{j_2}}{w(\mathrm{S}) - w_0}F + \cdots + \frac{w_{j_r}}{w(\mathrm{S}) - w_0}F \ge F/2,$$

and so

$$w_{j_1} + w_{j_2} + \cdots + w_{j_r} \ge \frac{w(\mathrm{S}) - w_0}{2}.$$

Let E_i be the event that more than $2F$ requests are served during block i. If event U_{i-1} and E_i occur, then there are at most $3F/2$ unserved type B requests in the beginning of block i (F arrived at the beginning of the block and there are at most $F/2$ from the previous block) but more than $2F$ requests were served. Therefore, at least one type A request was served during the block. Combining the results, if U_i, U_{i-1}, and E_i occur then:
- During block i at least $(w(\mathrm{S}) - w_0)/2$ time units were used for traveling between type B nodes.
- A Type A request was served during the block.

A block i is called *good* if the events U_i, U_{i-1}, and E_i occur. For any two (consecutive) good blocks the traveling cost is at least $(w(\mathrm{S}) - w_0)/2 + w_0 \ge w(\mathrm{S})/2$. Since none of the blocks satisfy Condition 1 or 2, it follows that for all i such that $\frac{1}{3}\sqrt{\frac{L}{w(\mathrm{S})}} \ge i \ge 1$ we have: $\Pr[U_i] \ge 3/4, \Pr[U_{i-1}] \ge 3/4$, and $\Pr[E_i] \ge 3/4$. Therefore:

$$\Pr[U_i \cap U_{i-1} \cap E_i] = 1 - \Pr[\neg(U_i \cap U_{i-1} \cap E_i)]$$
$$= 1 - \Pr[\neg U_i \cup \neg U_{i-1} \cup \neg E_i] \ge 1 - 1/4 - 1/4 - 1/4 = 1/4.$$

The sequence consists of $\frac{1}{3}\sqrt{\frac{L}{w(\mathrm{S})}}$ blocks. Therefore, the expected number of good blocks is $\frac{1}{4} \cdot \frac{1}{3}\sqrt{\frac{L}{w(\mathrm{S})}} = \frac{1}{12}\sqrt{\frac{L}{w(\mathrm{S})}}$ and of disjoint pairs of blocks is $\frac{1}{24}\sqrt{\frac{L}{w(\mathrm{S})}}$. Consequently, the expected number of lost requests is at least $\frac{1}{24}\sqrt{\frac{L}{w(\mathrm{S})}}\frac{w(\mathrm{S})}{2}$ and of served requests is:

$$E(\mathrm{ALG}(\sigma)) \le 3L - \frac{1}{48}w(\mathrm{S})\sqrt{\frac{L}{w(S)}} = 3L - \frac{1}{48}\left(\sqrt{w(\mathrm{S})L}\right).$$

- **We bound the performance of OPT′**: At each time unit until the final event, OPT′ serves the type A request that arrived at the same time unit. Consequently, from the final event until time unit $3L$, OPT′ serves the type A requests that arrived at the final event. Therefore, OPT′ serves $3L$ type A requests, and so OPT $\ge$ OPT′ $\ge 3L$.

The competitive ratio is

$$\frac{\text{OPT}(\sigma)}{E(\text{ALG}(\sigma))} \geq \frac{3L}{3L - \frac{1}{48}\left(\sqrt{w(\text{S})L}\right)} = 1 + \Omega\left(\sqrt{\delta}\right).$$

Note that in all 3 cases we get $1 + \Omega(\sqrt{\delta})$. This completes the proof. ∎

The following straightforward corollary improves the lower bound of $1 + \Omega(C/L)$ from [4]. Recall that n is the number of nodes in the metric space.

Corollary 1. *No deterministic or randomized online algorithm can achieve a competitive ratio better than $1 + \Omega\left(\sqrt{n/L}\right)$ when all traveling times takes one unit of time and $L \geq 256n$. Otherwise, if $L < 256n$, the bound becomes 1.00054.*

Proof. Let S be a star metric such that the weight of each edge is equal to $1/2$. Clearly, traveling between any two nodes requires one time unit and $w(\text{S}) = n/2$. Applying Theorem 1, we obtain the lower bound of $1 + \Omega\left(\sqrt{n/L}\right)$ (note that in this case $\delta = n/L$). ∎

Embedding of Metric Spaces. In this section we describe an embedding of a general metric space into a star metric with special properties. We begin by introducing some new definitions:

- We define $w(\text{T}) = \sum_{e \in V} w(e)$ for a rooted tree $\text{T} = (V, E)$, and let $P_\text{T}(v)$ denote the parent of node v in a rooted tree T.
- Let S be a star metric with a center c. We define $w_\text{S}(V) = \sum_{v \in V} w(c, v) = \sum_{v_i \in V} w_i$. It is clear that for a star S with leaves V, $w_\text{S}(V) = w(\text{S})$.
- Let $\text{T}_\text{G}(V)$ be the minimum weight connected tree that contains the set V (i.e., the minimum Steiner tree on these points) in the metric space G.

Recall that $MST(\text{G})$ denotes the weight of the minimal spanning tree (MST) in the metric space G.

Theorem 2. *For any given metric space* G *on nodes* V *and for any vertex* $v_0 \in V$ *there exists a star metric* S *with leaves* V *and an embedding* $f : \text{G} \to \text{S}$ *from* G *to* S *(f depends on v_0) such that:*

1. **Property 1:** $w(\text{S}) = MST(\text{G})$.
2. **Property 2:** *For every* $V' \subseteq V$ *such that* $v_0 \in V'$, $w(\text{T}_\text{G}(V')) \geq w_\text{S}(V')$.

Proof. We prove the theorem by describing a star metric that satisfies the required properties. Let G be a given metric space on nodes V with a vertex $v_0 \in V$. Let T be the MST for G created by applying Prims' algorithm with the root v_0. Let S be a star metric with leaves V such that for each $u \in V$, $w_u = w(u, P_\text{T}(u))$. Clearly, $w_{v_0} = 0$. We prove that S and v_0 satisfy the theorem's properties:

Property 1: Clearly, $w(\mathrm{S}) = w(\mathrm{T})$, and since T is a MST for G, $w(\mathrm{S}) = w(\mathrm{T}) = \mathrm{MST}(\mathrm{G})$.

Property 2: Assume by a contradiction that there exists $V' = \{v_0, v_{i_1}, \ldots, v_{i_r-1}\} \subseteq V$ such that $w(\mathrm{T_G}(V')) < w_S(V') = \sum_{j=1}^{r-1} w(v_{i_j}, P_\mathrm{T}(v_{i_j}))$.

Let $V'' = \{v_0, v_{i_1}, \ldots, v_{i_r-1}, \ldots, v_{i_k}\}$ be the vertices of $T_\mathrm{G}(V')$ (note that $V' \subseteq V''$). Consider the following process. Let $T' = T_\mathrm{G}(V')$. Run Prim's algorithm from node v_0. Each time Prim's adds a new node not in T', we add Prim's edge to T'. Note that Prim starts from node $v_0 \in T_\mathrm{G}(V')$ and we add each node not in T'. Hence, when Prim's algorithm finishes, T' is a tree on nodes V. Moreover, T' is T where edges $(v_{i_1}, P_\mathrm{T}(v_{i_1})), \ldots, (v_{i_k}, P_\mathrm{T}(v_{i_k}))$ were replaced by the edges of $T_\mathrm{G}(V')$. Since we assumed that $w(\mathrm{T_G}(V')) < \sum_{j=1}^{r-1} w(v_{i_j}, P_\mathrm{T}(v_{i_j}))$, and clearly $\sum_{j=1}^{r-1} w(v_{i_j}, P_\mathrm{T}(v_{i_j})) \leq \sum_{j=1}^{k} w(v_{i_j}, P_\mathrm{T}(v_{i_j}))$, we have $w(T') < w(T)$. This is a contradiction since T is an MST. ∎

Lower Bound for a General Metric Space. In this section we consider the case where the traveling time between nodes is represented by a metric space G. Note that a lower bound for a star metric space does not imply a lower bound for a general metric space. Recall that $\delta = \mathrm{TSP}(\mathrm{G})/L < 1$.

We use the embedding from Theorem 2 to prove a $1 + \Omega(\sqrt{\delta})$ lower bound.

Theorem 3. *No deterministic or randomized online algorithm can achieve a competitive ratio better than $1 + \Omega(\sqrt{\delta})$ for any given metric space G, when $\delta \leq \frac{1}{256}$. Otherwise, if $\delta > \frac{1}{256}$, the bound becomes 1.00054.*

3.2 Lower Bound for a Large Diameter Laxity Ratio (Case C)

In this section we consider the case where $L < \Delta(G)/2$ (recall that $\Delta(G)$ is the diameter and L is the laxity), and we show that the competitive ratio of any deterministic algorithm is unbounded.

Theorem 4. *No deterministic online algorithm can achieve a bounded competitive ratio for any metric space in which $L < \Delta(G)/2$.*

Proof. Let G be any metric space. Every $\Delta(G) + 1$ units of time we introduce a request with a laxity of L to a node which is at a distance of at least $\Delta(G)/2$ from the current location of the online algorithm (note that there is always such a node). It is clear that the algorithm can not serve any requests while OPT can serve all the requests. ∎

4 Upper Bounds

4.1 Asymptotically Optimal Algorithm for Case A

In this section we design a deterministic online algorithm, for a general metric space. The algorithm achieves a competitive ratio of $1 + o(1)$ when the minimum

In each phase $\ell = 1, 2, \ldots,$ do

- Beginning of the phase (at time $K(\ell - 1)$)
  - Decrease the deadline of each unserved request $([r, d], v)$ from d to $K \lfloor d/K \rfloor$.
  - Let R^ℓ be the collection of unserved requests such that their decreased deadline was not exceeded. Let S^ℓ be the K-length prefix of EDF (earliest deadline first) schedule (according to the modified deadline) of R^ℓ. Let $S_j^\ell \subseteq S^\ell$ denote the subset of requests at node v_j in S^ℓ.
    Let $v_{i_1}, v_{i_2}, \ldots, v_{i_n}$ denote the order of the nodes in the minimal TSP (or approximation).
  - ρ_ℓ consists of all requests of $S_{i_1}^\ell$ served consecutively, then all requests of $S_{i_2}^\ell$ served consecutively, and so on.
- During the phase (between time $K(\ell - 1)$ and time $K\ell$)
  - The requests are served according to ρ_ℓ (unserved requests in the suffix of ρ_ℓ, due to the end of the phase are dropped).

Fig. 1. Algorithm TSP-EDF.

laxity of the requests is asymptotically larger than the weight of the TSP (as shown in the previous sections, this is essential).

The algorithm is a natural extension of the BG algorithm from [4]. Our algorithm, which we call TSP-EDF, formally described in Fig. 1, works in phases of $K = \sqrt{\text{TSP}(G)L}$ time units. In each phase the algorithm serves requests node by node. The order of the nodes is determined by the minimum TSP or an approximation. The algorithm achieves a competitive ratio of $1 + O(\sqrt{\text{TSP}(G)/L})$ for $L > 10\text{TSP}(G)$.

Theorem 5. *The algorithm* TSP-EDF *attains a competitive ratio of* $1 + O(\sqrt{\text{TSP}(G)/L})$.

4.2 Constant Approximation Algorithm for Case B

In this section we design a deterministic online algorithm, for a general metric space where $L > 9\Delta(G)$ (recall that $\Delta(G)$ is the diameter of G). The algorithm achieves a constant competitive ratio. As shown in the previous section, no online algorithm can achieves a competitive ratio better 1.00054. A more precise analysis can replace $L > 9\Delta(G)$ with $L > (2 + \epsilon)\Delta(G)$ for any $\epsilon > 0$ and the approximation becomes $O(\frac{1}{\epsilon})$.

The algorithm which we call ORIENT-WINDOW (Fig. 2) combines the following ideas.

- The algorithm works in phase of $K = 3\Delta(G)$. In each phase the algorithm serves only requests that arrived in the previous phases, and will not expired during the phase. Due to this perturbation we lose a constant factor.
- The decision which requests will be served in a phase ignore their deadlines. Due to this violation of EDF we lose a constant factor.

In each phase $\ell = 1, 2, \ldots$, do

- Beginning of the phase (at time $K(\ell - 1)$)
  - Decrease the deadline of each unserved request $([r, d], v)$ from d to $K\lfloor d/K \rfloor$.
  - Let R^ℓ be the collection of unserved requests such that their decreased deadline was not exceeded. Let $S_j^\ell \subseteq R^\ell$ denote the subset of requests at node v_j in R^ℓ.
  - Using a constant approximation algorithm solve the unrooted orienteering problem with budget $\Delta(G)$ where the prize of a node v_j is the number of requests in S_j^ℓ. Let $v_{i_1}, v_{i_2}, \ldots, v_{i_r}$ denote the order of the nodes in the solution.
  - ρ_ℓ consists of all requests of $S_{i_1}^\ell$ scheduled consecutively, then all requests of $S_{i_2}^\ell$ scheduled consecutively, and so on.
- During the phase (between time $K(\ell - 1)$ and time $K\ell$)
  - The requests are served according to ρ_ℓ (unserved requests in the suffix of ρ_ℓ, due to the end of the phase are dropped).

Fig. 2. Algorithm ORIENT-WINDOW

– In each phase the algorithm serves requests node by node. The order of the nodes is determined by solving an orienteering problem. Since a constant approximation algorithm is known to the orienteering problem, we lose a constant factor.

Theorem 6. *The algorithm ORIENT-WINDOW attains a competitive ratio of* $O(1)$.

References

1. Arkin, E.M., Mitchell, J.S., Narasimhan, G.: Resource-constrained geometric network optimization. In: SOCG, pp. 307–316. ACM (1998)
2. Ausiello, G., Feuerstein, E., Leonardi, S., Stougie, L., Talamo, M.: Algorithms for the on-line travelling salesman. Eindhoven University of Technology, Department of Mathematics and Computing Sciences (1999)
3. Awerbuch, B., Azar, Y., Blum, A., Vempala, S.: New approximation guarantees for minimum-weight k-trees and prize-collecting salesmen. SIAM J. Comput. **28**(1), 254–262 (1998)
4. Azar, Y., Feige, U., Gamzu, I., Moscibroda, T., Raghavendra, P.: Buffer management for colored packets with deadlines. In: SPAA 2009, pp. 319–327. ACM (2009)
5. Bansal, N., Blum, A., Chawla, S., Meyerson, A.: Approximation algorithms for Deadline-TSP and vehicle routing with time-windows. In: STOC, pp. 166–174. ACM (2004)
6. Bar-Yehuda, R., Even, G., Shahar, S.M.: On approximating a geometric prize-collecting traveling salesman problem with time windows. J. Algorithms **55**(1), 76–92 (2005)
7. Blum, A., Chawla, S., Karger, D.R., Lane, T., Meyerson, A., Minkoff, M.: Approximation algorithms for orienteering and discounted-reward TSP. In: Proceedings of the 44th Annual IEEE Symposium on Foundations of Computer Science, pp. 46–55. IEEE (2003)
8. Chekuri, C., Korula, N.: Approximation algorithms for orienteering with time windows. arXiv preprint arXiv:0711.4825 (2007)

9. Chekuri, C., Korula, N., Pál, M.: Improved algorithms for orienteering and related problems. ACM Trans. Algorithms (TALG) **8**(3), 23 (2012)
10. Chekuri, C., Kumar, A.: Maximum coverage problem with group budget constraints and applications. In: Jansen, K., Khanna, S., Rolim, J.D.P., Ron, D. (eds.) APPROX/RANDOM -2004. LNCS, vol. 3122, pp. 72–83. Springer, Heidelberg (2004). doi:10.1007/978-3-540-27821-4_7
11. Desrochers, M., Desrosiers, J., Solomon, M.: A new optimization algorithm for the vehicle routing problem with time windows. Oper. Res. **40**(2), 342–354 (1992)
12. Desrochers, M., Lenstra, J.K., Savelsbergh, M.W., Soumis, F.: Vehicle routing with time windows: optimization and approximation. Veh. Routing: Methods Stud. **16**, 65–84 (1988)
13. Feuerstein, E., Stougie, L.: On-line single-server dial-a-ride problems. Theoret. Comput. Sci. **268**(1), 91–105 (2001)
14. Golden, B.L., Levy, L., Vohra, R.: The orienteering problem. Naval Res. Logist. **34**(3), 307–318 (1987)
15. Irani, S., Lu, X., Regan, A.: On-line algorithms for the dynamic traveling repair problem. J. Sched. **7**(3), 243–258 (2004)
16. Jaillet, P., Lu, X.: Online traveling salesman problems with rejection options. Networks **64**(2), 84–95 (2014)
17. Karuno, Y., Nagamochi, H.: 2-approximation algorithms for the multi-vehicle scheduling problem on a path with release and handling times. Discrete Appl. Math. **129**(2), 433–447 (2003)
18. Krumke, S.O., Megow, N., Vredeveld, T.: How to whack moles. In: Solis-Oba, R., Jansen, K. (eds.) WAOA 2003. LNCS, vol. 2909, pp. 192–205. Springer, Heidelberg (2004). doi:10.1007/978-3-540-24592-6_15
19. Nagamochi, H., Ohnishi, T.: Approximating a vehicle scheduling problem with time windows and handling times. Theoret. Comput. Sci. **393**(1), 133–146 (2008)
20. Savelsbergh, M.W.: Local search in routing problems with time windows. Ann. Oper. Res. **4**(1), 285–305 (1985)
21. Tan, K.C., Lee, L.H., Zhu, Q., Ou, K.: Heuristic methods for vehicle routing problem with time windows. Artif. Intell. Eng. **15**(3), 281–295 (2001)
22. Thangiah, S.R.: Vehicle Routing with Time Windows Using Genetic Algorithms. Citeseer (1993)
23. Tsitsiklis, J.N.: Special cases of traveling salesman and repairman problems with time windows. Networks **22**(3), 263–282 (1992)

A PTAS for the Cluster Editing Problem
on Planar Graphs

André Berger, Alexander Grigoriev$^{(\boxtimes)}$, and Andrej Winokurow

Maastricht University School of Business and Economics,
P.O. Box 616, 6200 MD Maastricht, The Netherlands
{a.berger,a.grigoriev,a.winokurow}@maastrichtuniversity.nl

Abstract. The goal of the cluster editing problem is to add or delete a minimum number of edges from a given graph, so that the resulting graph becomes a union of disjoint cliques. The cluster editing problem is closely related to correlation clustering and has applications, e.g. in image segmentation. For general graphs this problem is APX-hard. In this paper we present an efficient polynomial time approximation scheme for the cluster editing problem on graphs embeddable in the plane with a few edge crossings. The running time of the algorithm is $2^{O\left(\epsilon^{-1}\log(\epsilon^{-1})\right)}n$ for planar graphs and $2^{O\left(k^2\epsilon^{-1}\log\left(k^2\epsilon^{-1}\right)\right)}n$ for planar graphs with at most k crossings.

Keywords: Graph approximation · Correlation clustering · Cluster editing · PTAS · k-planarity · Microscopy cell segmentation

1 Introduction

The task of the *cluster editing problem* is to find a minimum number of edges to be deleted or added to an input graph G to achieve a union of disjoint complete connected components G^*. The resulting graph G^* is the solution to the problem.

Different definitions and interpretations of this problem were first discussed in [5,16,23]. More recently, motivated by machine learning problems concerning document classification, Bansal et al. [4] reintroduced the same concept under the name *correlation clustering*. The reformulation of the cluster editing problem as correlation clustering motivated computer scientists to apply cluster editing/correlation clustering in the field of image segmentation [17,22]. In turn, this triggered mathematicians to start looking for efficient algorithms able to attack the problem. The cluster editing problem is NP-hard and even APX-hard [8]. Ageev et al. [1] presented a polynomial time approximation scheme (PTAS) for the case of at most two clusters. In [15], Il'ev et al. present a PTAS for a fixed number of clusters and graphs with sub-quadratic number of edges. Recently, Klein et al. [18] introduced a PTAS for the planar two-edge-connected augmentation problem, which is equivalent to the correlation clustering problem on planar graphs.

Parameterized complexity of the cluster editing problem was also studied intensively, e.g., with the number of modified edges being a parameter. The currently fastest exact algorithm for this problem is by Böcker et al. [6] and

© Springer International Publishing AG 2017
K. Jansen and M. Mastrolilli (Eds.): WAOA 2016, LNCS 10138, pp. 27–39, 2017.
DOI: 10.1007/978-3-319-51741-4_3

it runs in $O\left(1.82^{k}+n^{3}\right)$ time, where k is the number of modified edges and n is the number of vertices in the input graph. To the best of our knowledge, the cluster editing problem parameterized by the treewidth was not considered before. In Sect. 4, we prove that the cluster editing problem parameterized by treewidth ω belongs to $\mathbb{FPT}$ and can be solved in time $2^{O(\omega \log \omega)}n$.

In the original formulation the cluster editing problem and correlation clustering problem are equivalent. In correlation clustering some edges of a complete graph are labeled with $<->$ and the rest with $<+>$. If in the solution graph two vertices are in the same cluster and the edge between them is positive or two vertices are in different clusters and the edge is negative, then it is an *agreement*. In both other cases we have a *disagreement*. The goal is to partition the vertices in such a way, that the total number of disagreements is minimized.

The correlation clustering problem was generalized to non-complete graphs by Klein et al. [18]. In that paper, the non-existing edges are introduced as neutral, so that they create neither agreements nor disagreements. In contrast to the correlation clustering problem, cluster editing on planar graphs is a restriction of the original problem. An example of an input instance for the planar correlation clustering problem is shown in Fig. 1a. Here, the optimal correlation clustering splits the graph in four connected components and the disagreements correspond to the red edges. The optimal solution to the correlation clustering has fewer clusters than the solution of the equivalent cluster editing problem in the majority of the cases (Figs. 1b, c).

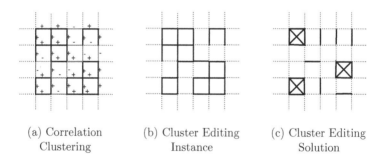

(a) Correlation Clustering   (b) Cluster Editing Instance   (c) Cluster Editing Solution

Fig. 1. Graph example demonstrating the difference between the planar correlation clustering problem and the planar cluster editing problem

Since the idea of considering planar graphs for cluster editing and correlation clustering was originally based on its application in image segmentation, let us recall the basic idea of image segmentation. The goal of image segmentation is to partition the pixels of an image in such a way that the discovered clusters correspond to certain objects located on that image. Every image can be considered as a graph of pixels, where two pixels are connected if they are likely to be in the same cluster.

In microscopy cell segmentation [10] and in dense crowd counting [14] the sizes of the clusters are small, which means that the distance between any two

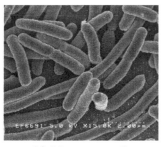

(a) Microscopy cell
segmentation

(b) People counting in
dense crowds

Fig. 2. Image segmentation examples with small clusters

connected pixels is bounded (Fig. 2). This results in graphs embeddable in the plane with a few crossings per edge. Following these motivating applications, in this paper we consider the cluster editing problem restricted to these, nearly planar graphs. The outline of the paper is as follows. We design an exact algorithm for graphs of bounded treewidth in Sect. 4 and a PTAS for planar graphs based on the Baker's technique [3] in Sect. 5. In the last section we show that the introduced algorithm also works for graphs embeddable with a few edge crossings per edge.

2   Definitions and Notations

We first review some basic notation and definitions, including the definition of treewidth, tree decomposition and nice tree decomposition.

Let G be a simple graph. We denote the set of vertices of G by $V[G]$ and the set of edges by $E[G]$. Let n be the number of vertices in G. For a subset of vertices $W \subset V[G]$ the subgraph induced by W is denoted by $G[W]$. Let $\delta_G(W_1, W_2) \subset V[G]^2$ be the set of edges connecting disjoint sets $W_1 \subset V[G]$ and $W_2 \subset V[G]$. For any $v \in V[G]$ the degree of a vertex v in the graph G is denoted by $\deg_G(v)$. The neighborhood of a vertex $v \in V[G]$ in a graph G is denoted by $N_G(v)$ and is defined as $\{w \in V[G] \mid \{v, w\} \in E[G]\}$.

Definition 1 (Tree Decomposition). *A tree decomposition of a graph $G = (V, E)$ is a pair $(\{X_i : i \in I\}, T = (I, F))$, where $\{X_i : i \in I\}$ is a family of so-called bags, which are subsets of V and T is a tree, satisfying:*

1. *$\bigcup_{i \in I} X_i = V$,*
2. *for any edge $\{u, v\} \in E$ there exists an $i \in I$ such that $u, v \in X_i$, and*
3. *for any $v \in V$, $\{i \in I : v \in X_i\}$ induces a subtree of T.*

The width of the tree decomposition is $\max_{i \in I} |X_i| - 1$. The treewidth $\omega(G)$ (or simply ω) of graph G is the minimum width of tree decompositions of G.

Definition 2. *A tree decomposition* $(\{X_i : i \in I\}, T)$ *of a graph* G *rooted at some* $r \in I$ *is called nice, if the following conditions are satisfied:*

1. *Every node of the tree* T *has at most two children.*
2. *If a node* $i \in I$ *has two children* j *and* k, *then* $X_i = X_j = X_k$. *In this case* i *is called a join node.*
3. *If a node* i *has one child* j, *then either*
   (a) $X_i = X_j \cup \{w\}$ *for some* $w \in V \backslash X_j$ (*i is an insert node*), *or*
   (b) $X_i = X_j \backslash \{w\}$ *for some* $w \in X_j$ (*i is called a forget node*).

Lemma 1 *[19, Lemma 13.1.3]. Given a tree decomposition of a graph* G *on* $O(n)$ *nodes and of width* w, *we can find a nice tree decomposition of* G *on* $O(n)$ *nodes and having the same width* w *in* $O(wn)$ *time.*

3   Cluster Editing

Given a graph $G = (V, E)$, let $\mathcal{M}(V)$ be the family of all graphs with vertex set V and all connected components being cliques.

Definition 3. *For a given graph* G, *the cluster editing problem is the problem of finding a graph* $G^* = (V, E^*) \in \mathcal{M}(V)$ *that minimizes*

$$\rho(G, G^*) = |E^* \backslash E| + |E \backslash E^*|.$$

Here, $\rho(G, G^*)$ is the number of edges in G we have to modify, i.e., add or delete, to get G^*. Later in this paper we consider the graph G being planar. Notice however, this does not require the solution graph G^* to be planar.

Lemma 2. *For any connected component* C *of* G^* *and for any subset* $C_1 \subset C$, *we have that*

$$\delta_G(C_1, C \backslash C_1) \geq \frac{1}{2}|C_1||C \backslash C_1|.$$

Proof. Let H^* be a subgraph of G^* obtained by removing all edges between C_1 and $C_2 := C \backslash C_1$. Obviously, $H^* \in \mathcal{M}(V)$ and since G^* is optimal

$$\rho(G, G^*) \leq \rho(G, H^*). \tag{1}$$

Here $\rho(G, G^*)$ and $\rho(G, H^*)$ can be decomposed as follows

$$\rho(G, G^*) = \rho(G[C_1], G^*[C_1]) + \rho(G[C_2], G^*[C_2]) + \rho(G[V \backslash C], G^*[V \backslash C])$$
$$+ (|\delta_{G^*}(C_1, C_2)| - |\delta_G(C_1, C_2)|) + (|\delta_G(C, V \backslash C)| - |\delta_{G^*}(C, V \backslash C)|),$$
$$\rho(G, H^*) = \rho(G[C_1], H^*[C_1]) + \rho(G[C_2], H^*[C_2]) + \rho(G[V \backslash C], H^*[V \backslash C])$$
$$+ (|\delta_G(C_1, C_2)| - |\delta_{H^*}(C_1, C_2)|) + (|\delta_G(C, V \backslash C)| - |\delta_{H^*}(C, V \backslash C)|).$$

Substituting $\rho(G, G^*)$ and $\rho(G, H^*)$ in (1) proves that

$$|\delta_G(C_1, C_2)| \geq \frac{1}{2}|C_1||C_2|.$$

Lemma 3. *For any connected component C of G^**

(a) $\deg_{G[C]}(v) \geq \frac{1}{2}(|C| - 1)$, *for any vertex $v \in C$,*
(b) $\deg_{G[C]}(v) + \deg_{G[C]}(w) \geq |C|$, *for any edge $\{v, w\} \in E[G[C]]$,*
(c) $\deg_{G[C]}(v) + \deg_{G[C]}(w) + \deg_{G[C]}(u) \geq \frac{3}{2}(|C| + 1)$, *for any $v, u, w \in C$, s.t.*
    $\{v, w\}, \{v, u\}, \{u, w\} \in E[G[C]]$,
(d) *If G is planar, then $|C| \leq 8$.*

Proof. The first three inequalities follow directly from Lemma 2 by setting $C_1 = \{v\}$, $C_1 = \{v, w\}$ and $C_1 = \{v, w, u\}$.

To prove the last point we use the well known corollary of Euler's formula for planar graphs: $|E| \leq 3|V| - 6$. Since $G[C]$ is planar, there is a vertex of degree at most 5 in $G[C]$. Therefore, by part *(a)* $|C| \leq 11$. Using Euler's formula again we can prove that there is a vertex of degree at most 4, and by part *(a)* $|C| \leq 9$. Now, let us assume that $|C| = 9$. By part *(a)* there is no vertex of degree smaller than 4. So, there is a vertex v_0 of degree 4. Its neighboring vertices v_1, v_2, v_3, v_4 have degree at least 5 by part *(b)* of this corollary. W.l.o.g., by part *(c)*, v_1 and at least one of v_2, v_3, v_4 have degree at least 6. Summing up all the degrees of the vertices we get at least $4 + 6 + 6 + 5 + 5 + 4 + 4 + 4 + 4 = 42$, which is the maximum degree according to Euler's formula. Hence, v_1 and v_3 have degree 6, v_2, v_3 have degree 5 and the remaining vertices have degree 4. So, there should be an edge between 2 vertices of degree 4, which is a contradiction to part *(b)*.

Lemma 4. *For a connected component C of G and $v \in C$, $N_{G^*}(v) \subset C$.*

Proof. Let C_1 be the connected component in G^* containing v and $C_1 \backslash C \neq \emptyset$, then subgraph H^* of G^*, where the edges in $\delta(C_1 \cap C, C_1 \backslash C)$ are deleted, has lower objective value. This contradicts G^* being optimal.

Alternatively, Lemma 4 states, that if G has several connected components, then we can solve the problem separately on each of the components and then combine the solutions.

4   Linear Time Algorithm for Graphs of Bounded Treewidth

In this section we introduce an algorithm which solves the cluster editing problem for any input graph $G = (V, E)$ of bounded treewidth w in linear time. We assume that G is connected, otherwise by Lemma 4 we can consider each connected component separately.

Let $F \subset V^2$ be the set of edges we modify in G to obtain G^*, then $\rho(G, G^*) = |F|$. The following monadic second order formula decides if $G^* \in \mathcal{M}(V)$.

$$\forall x, y, z \in V \ [y = z \lor \forall e, f \in E \Delta F \ y, x \in e \land x, z \in f \implies \exists h \in E \Delta F \ y, z \in h],$$

where $E \Delta F$ means the symmetric difference between two sets E and F, i.e., $E \Delta F = E \backslash F \cup F \backslash E$. This formula is true if and only if in the graph G^* every path of length two is a part of a cycle of length three. Since the length of the formula is constant, by the optimization version of Courcelle's theorem [2,9] we can find $G^* \in \mathcal{M}(V)$ that minimizes $\rho(G, G^*)$ in time $O\left(f(\omega) \cdot n\right)$, where $f(\cdot)$ is a function which doesn't depend on the input graph G and ω is the treewidth.

The optimization version of Courcelle's theorem [2,9] shows that for a graph of bounded treewidth we can solve the cluster editing problem in linear time. Unfortunately, it neither gives an implementable combinatorial algorithm for the cluster editing problem of bounded treewidth nor a precise bound for the running time. Therefore, in the second part of this section we introduce an algorithm based on tree decomposition and dynamic programming with running time $2^{O(\omega \log \omega)} n$.

Lemma 5. *Any connected component C of the solution graph G^* consists of at most $4\omega - 1$ vertices, where ω is the treewidth of the graph $G[C]$.*

Proof. By Rose [21], the number of edges in the graph $G[C]$ of treewidth ω is at most $\omega |C| - \omega(\omega + 1)/2$. Therefore, the minimal vertex degree in $G[C]$ is at most $2\omega - 1$. By part (a) of Lemma 3, $|C| \le 4\omega - 1$.

Suppose a nice tree decomposition $(\{X_i \mid i \in I\} \mid T = (I, F))$ of width w on $O(n)$ nodes is given. For each $i \in I$ define $G_i = (V_i, E_i)$ as a subgraph of G induced by all vertices in bag i and all bags below i in the tree T.

For a set $F \in \mathcal{F}_i = \{F \subset X_i \times X_i \mid (X_i, E[G[X_i]] \Delta F) \in \mathcal{M}(X_i)\}$ and a function $S \in \mathcal{S}_i = \{S : X_i \to \{0, \ldots, 4\omega - 2\}\}$ we restrict the cluster editing problem on the graph G_i to the cluster editing problem where the subgraph induced by X_i from the solution graph is $(X_i, E[G[X_i]] \Delta F)$ and for each $v \in X_i$ the size of a clique containing v in the solution is $S(v) + \left|N_{G_i^*[X_i]}(v)\right| + 1$. In other words, $S(v)$ is the number of neighbors of v in the solution graph of the restricted problem outside the bag X_i. We refer to the solution of this restricted problem as $G_i^*(S, F)$. Let $f_i(S, F) = \rho(G_i, G_i^*(S, F))$.

Theorem 1. *The cluster editing problem can be solved in time $2^{O(\omega \log \omega)} n$.*

Proof. For any $i \in I, S \in \mathcal{S}_i, F \in \mathcal{F}_i$ we compute $G_i^*(S, F)$ and $f_i(S, F)$ using dynamic programming. We start from the leafs of the nice tree decomposition (X, T) and move towards the root by considering i being a leaf, a join node, an insert node or a forget node. Hence, when computing $G_i^*(S, F)$ we may assume that for every (not necessarily immediate) successor node j of i in T, $f_j(S_j, F_j)$ and $G_j^*(S_j, F_j)$ are known for any $F_j \in \mathcal{F}_j$ and $S_j \in \mathcal{S}_j$.

Leaf. In this case $G_i = G[X_i]$. Therefore, F uniquely defines the solution $G_i^*(S, F) = (X_i, E[G[X_i]] \Delta F)$ and

$$f_i(S, F) = \begin{cases} |F|, \text{ if } G_i^*(S, F) \in \mathcal{M}(X_i), \\ \infty, \text{ otherwise.} \end{cases}$$

Join Node. If i is a join node, then $X_i = X_j = X_k$, where j and k are children of the node i in the tree T. Since the subgraph of the solution induced by X_i is fixed by F, $G_i^*(S, F)$ is a combination of $G_j^*(S_j, F)$ and $G_k^*(S_k, F)$, where S is split into S_j and S_k to minimize $f_j(S_j, F) + f_k(S_k, F)$, i.e.

$$f_i(S, F) = \left(\min_{\substack{S_j, S_k \in \mathcal{S}_i: \\ \forall v \in X_i \ S_j(v) + S_k(v) = S(v)}} [f_j(S_j, F) + f_k(S_k, F)] \right) - |F|.$$

Here, we subtract $|F|$ as otherwise edges of F are counted twice.

Insert Node. Let i be an insert node, i.e. $X_i = X_j \cup \{w\}$, where $w \notin X_j$. Since $X_j \subset X_i$, a subgraph induced by X_j from the solution is uniquely defined by F, moreover, $N_{G_i^*(S,F)}(w) \cap X_j$ is known. Let $F_j = F_i \cap (X_j \times X_j)$ and for each $u \in X_j$ let $S_j(u) = S(u)$.

All edges modified in G_i to get $G_i^*(S, F)$ are either in $V[G_j] \times V[G_j]$ or are incident to w. The number of the first ones is $f_j(S_j, F_j)$. It remains to find the number of edges incident to w to be deleted (the right upper corner on Fig. 3) or to be added (the left lower corner on Fig. 3) in G_i to get $G_i^*(S, F)$.

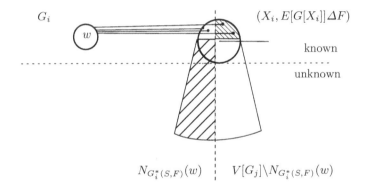

Fig. 3. Insert node w.

In the optimal solution $G_i^*(S, F)$ w is connected to $N_{G_i^*(S,F)}(w) \cap X_j$ and $N_{G_i^*(S,F)}(w) \backslash X_j$. In the original graph G vertices of the first set are either connected to w or not, where vertices of the second set are never connected to w, because of the tree decomposition properties. So, we have to add all missing edges from w to $N_{G_i^*(S,F)}(w) \cap X_j$ (we know all these edges) and also all edges from w to $N_{G_i^*(S,F)}(w) \backslash X_j$ (the number of these edges is $S(w)$). We also have to delete all edges from w to $X_j \backslash N_{G_i^*(S,F)}(w)$ (we again know all these edges). We do not have to delete edges from w to $V[G_j] \backslash (X_j \backslash N_{G_i^*(S,F)}(w))$, because they are not existing in graph G since w and $V[G_j] \backslash (X_j \backslash N_{G_i^*(S,F)}(w))$ lie on different sides of the bag X_j. Hence, knowing S and F we can calculate the number $f_i(S, F)$ of edges to be modified. Then, we reconstruct $G_i^*(S, F)$.

Forget Node. Let i be a forget node, i.e., for any successor node j of i, $X_j = X_i \cup \{w\}$, where $w \notin X_i$. Since $G_i = G_j$,

$$f_i(S, F) = f_j(S_j, F_j)$$

for a proper choice of $S_j \in \mathcal{S}_j, F_j \in \mathcal{F}_j$. Here, S_j and F_j depend on how w is connected to $V[G_i]$.

Assume that in the optimal solution $G_i^*(S, F)$ vertex w is not connected to any vertex in X_i, and it is connected to exactly $t \in \{0, \ldots 4\omega - 2\}$ vertices outside of X_i, then

$$F_j = F \text{ and } S_j(u) = \begin{cases} S(u), & u \neq w, \\ t, & \text{otherwise} \end{cases}$$

Otherwise, if in the optimal solution $G_i^*(S, F)$ vertex w is connected to some $v \in X_i$ and also to exactly $t \in \{0, \ldots 4\omega - 3\}$ vertices outside of X_i, then w is also connected to $N_{G_i^*(S,F)}(v) \cap X_i$. Therefore,

$$F_j = F \cup \{\{w, u\} \mid u \in (\{v\} \cup N_{G_i^*(S,F)}(v) \cap X_i) \Delta (N_G(w) \cap X_i))\}$$

$$S_j(u) = \begin{cases} S(u) - 1, & u \in N_{G_i^*(S,F)}(v) \vee u = v \wedge u \neq w, \\ S(u), & u \notin N_{G_i^*(S,F)}(v) \wedge u \neq w, \\ t, & u = w. \end{cases}$$

Out of all possibilities for $N_{G_i^*(S,F)}(w) \cap X_i$ and $t \in \{0, \ldots 4\omega - 2\}$ the minimal $f_j(S_j, F_j)$ is equal to $f_i(S, F)$. After we found $f_i(S, F)$ we reconstruct $G_i^*(S, F)$.

Root. The optimal solution G^* for graph G can be found by enumerating of all possible $S \in \mathcal{S}_r, F \in \mathcal{F}_r$ and selecting the one minimizing $f_r(S, F)$, i.e., $\rho(G, G^*) = \min_{S \in \mathcal{S}_r, F \in \mathcal{F}_r} f_r(S, F)$.

Following this scheme we find $f_i(S, F)$ and $G_i^*(S, F)$ for every $S \in \mathcal{S}_i, F \in \mathcal{F}_i$ and $i \in I$ in time $O\left((4\omega - 1)^{\omega+1}\right)$. Therefore, the overall time complexity is $O\left(\max_{i \in I} |\mathcal{S}_i| \max_{i \in I} |\mathcal{F}_i| (4\omega - 1)^{\omega+1} n\right)$. Here, $\max_{i \in I} |\mathcal{S}_i| = O\left((4\omega - 1)^{\omega+1}\right)$ and $\max_{i \in I} |\mathcal{F}_i| = O\left((\omega + 1)^{(\omega+1)}\right)$ and therefore the overall running time is $O\left((\omega + 1)^{(\omega+1)}(4\omega - 1)^{2(\omega+1)} n\right)$.

5   Baker's Technique for Planar Graphs

The idea of Baker [3] is to divide a given planar graph into independent components of bounded treewidth using outerplanarity. Having solved a problem on the components, an approximate solution for the original instance can be found by merging solutions together. In this way Baker's technique is used to design a polynomial time approximation schemes.

In this section the input graph is assumed to be planar and connected. If the input graph is not connected, then by Lemma 4 the connected components can be considered separately. For a given planar embedding of a graph $G = (V, E)$

we recursively assign vertices to levels. A vertex is assigned to level 1, if it is incident to the outer face of the embedding. A vertex v is assigned to level i, denoted by $v \in V_i$, if after removing levels from 1 through $i-1$, the vertex v is incident to the outer face.

For any $d \in \mathbb{N}$ and $l \in \{0, \ldots, d-1\}$ we define $E_{d,l} \subset V \times V$ as all edges between consecutive layers $l \pmod d$ and $l+1 \pmod d$, i.e. $E_{d,l} = \bigcup_{s \in \mathbb{N}} (V_{l+sd} \times V_{l+1+sd})$. Further, let $H_{d,l} = (V, E \backslash E_{d,l})$ be the subgraph of G achieved by deleting all edges between consecutive layers $l \pmod d$ and $l+1 \pmod d$ and let $H_{d,l}^*$ be the optimal solution of the cluster editing problem on the graph $H_{d,l}$ (Fig. 4). By Lemma 4, $E[H_{d,l}^*] \cap E_{d,l} = \emptyset$.

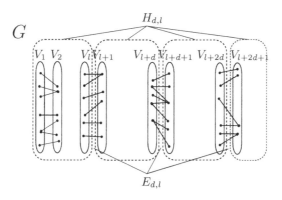

Fig. 4. Splitting the graph G in layers.

Theorem 2. *For $d_\epsilon = \lceil 48/\epsilon \rceil$ and $l_\epsilon \in \mathrm{argmin}_{l \in \{0, \ldots, d_\epsilon - 1\}} |E[G] \cap E_{d,l}|$, the graph $H_{d_\epsilon, l_\epsilon}^*$ is an $(1+\epsilon)$-approximation of the optimal solution G^* and it can be computed in time $2^{O(\epsilon^{-1} \log(\epsilon^{-1}))} n$.*

Proof. Since G^* is a solution to the cluster editing problem on the graph $H_{d,l}$, for any $d \in \mathbb{N}$ and $l \in \{0, \ldots, d-1\}$, we have that

$$\rho(H_{d,l}, H_{d,l}^*) \le \rho(H_{d,l}, G^*) \stackrel{\mathrm{def}}{=} |E[H_{d,l}] \backslash E[G^*]| + |E[G^*] \backslash E[H_{d,l}]|$$
$$\stackrel{\mathrm{Fig.\,5a}}{=} |E \backslash E_{d,l} \backslash E[G^*]| + |E[G^*] \backslash E| + |E[G^*] \cap E \cap E_{d,l}|$$
$$= |E \backslash E[G^*]| - |E \cap E_{d,l} \backslash E[G^*]| + |E[G^*] \backslash E| + |E[G^*] \cap E \cap E_{d,l}|$$
$$= \rho(G, G^*) - |E \cap E_{d,l} \backslash E[G^*]| + |E \cap E_{d,l} \cap E[G^*]|. \qquad (2)$$

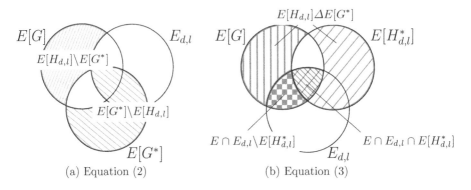

Fig. 5. Proofs of Eqs. (2) and (3)

Then, since $H_{d,l} = (V, E \setminus E_{d,l})$,

$$\rho(G, H_{d,l}^*) \overset{\text{Fig. 5b}}{=} \rho(H_{d,l}, H_{d,l}^*) - \left| E \cap E_{d,l} \cap E[H_{d,l}^*] \right| + \left| E \cap E_{d,l} \setminus E[H_{d,l}^*] \right| \quad (3)$$

$$\overset{(2)}{\leq} \rho(G, G^*) - \left| E \cap E_{d,l} \setminus E[G^*] \right| + \left| E \cap E_{d,l} \cap E[G^*] \right|$$
$$- \left| E \cap E_{d,l} \cap E[H_{d,l}^*] \right| + \left| E \cap E_{d,l} \setminus E[H_{d,l}^*] \right|$$
$$\leq \rho(G, G^*) + 2 \left| E \cap E_{d,l} \cap E[G^*] \right| \quad (4)$$

Since the graph G is connected, to cut it in K cliques we have to delete at least $K - 1$ edges. If we have to delete exactly $K - 1$ edges, by Lemma 3 the graph G consists of K cliques of at most 8 vertices, each connected with $K - 1$ edges in a chain similar graph. The treewidth of such a graph is at most 8 and, therefore, by Theorem 1, the exact solution can be found in $O(n)$ time.

Now, assume that at least K edges have to be deleted. Moreover, since by Lemma 3 the largest clique has at most 8 vertices, $K \geq |V|/8$. Further, since G is planar, $|E| \leq 3|V| - 6$. Combining these facts, we get

$$\rho(G, G^*) \geq K \geq |V|/8 \geq |E|/24. \quad (5)$$

Thus, for $d_\epsilon = \lceil 48/\epsilon \rceil$ and $l_\epsilon \in \operatorname{argmin}_{l \in \{0, \dots, d-1\}} |E[G] \cap E_{d,l}|$, we derive

$$\rho(G, H_{d_\epsilon, l_\epsilon}^*) \overset{(4)}{\leq} \rho(G, G^*) + 2|E[G] \cap E_{d_\epsilon, l_\epsilon}| \leq \rho(G, G^*) + 2|E[G]|/d_\epsilon \quad (6)$$
$$\leq (1 + 48/d_\epsilon)\rho(G, G^*) \leq (1 + \epsilon)\rho(G, G^*)$$

and $H_{d_\epsilon, l_\epsilon}^*$ is an $(1 + \epsilon)$-approximation of the original problem.

It remains to prove, that $H_{d_\epsilon, l_\epsilon}^*$ can be computed in time $2^{O(\epsilon^{-1} \log(\epsilon^{-1}))} n$. Since the diameter of the planar graph $H_{d,l}$ is bounded by d we can find a tree decomposition of treewidth $3d - 2$ in time $O(dn)$ [13]. From this tree decomposition we can construct in $O(dn)$-time a nice tree decomposition with

$O(n)$ nodes and of treewidth $3d - 2$ [19]. Having a nice tree decomposition of width at most $3d - 2$ by Theorem 1, $H^*_{d_\epsilon, l_\epsilon}$ can be computed in time $2^{O((3d-2)\log(3d-2))} n = 2^{O(\epsilon^{-1}\log(\epsilon^{-1}))} n$.

6   Cluster Editing on Graphs with a Few Crossings per Edge

A graph is called k-*planar* if it has a planar embedding, where each edge is crossed at most k times. Image segmentation often deals with k-planar graphs. For instance, if the furthest connected pixels are on a Chebyshev distance of at most R, then each edge is crossed at most $(2R)^2$ times. Hence, for such applications the input graph for the cluster editing problem is $(2R)^2$-planar.

Let us first discuss some properties of k-planar graphs. Using the results of Dujmović et al. [11,12], it can be easily shown that a k-planar graph with diameter d has treewidth at most $12kd + 6k + 2d$.

Lemma 6. *If the input graph G is k-planar for $k > 0$, then any connected component C of the cluster editing solution G^* has at most $18\sqrt{k}$ vertices.*

Proof. The number of edges in a k-planar graph with n vertices and $k > 0$ is at most $4.108\sqrt{k}n$ [20]. Therefore, the minimal degree is at most $8.216\sqrt{k}$ and by part (a) of Lemma 3, $|C| \leq 16.432\sqrt{k} + 1 \leq 18\sqrt{k}$.

Theorem 3. *For $d_\epsilon = \lceil 180k/\epsilon \rceil$ and $l_\epsilon \in \operatorname{argmin}_{l \in \{0,\ldots,d_\epsilon - 1\}} |E[G] \cap E_{d,l}|$, the graph $H^*_{d_\epsilon, l_\epsilon}$ is an $(1 + \epsilon)$-approximation of the optimal solution G^* and it can be computed in time $2^{O(k^2 \epsilon^{-1} \log(k^2 \epsilon^{-1}))} n$.*

Proof. The proof is similar to the proof of Theorem 2. If the treewidth of the graph G is at most $18\sqrt{k}$, then by Theorem 1, the cluster editing problem on the graph G can be solved exactly in $2^{O(\sqrt{k}\log\sqrt{k})} n$ time. Otherwise, similar to (5) we prove that $\rho(G, G^*) \geq K \geq \frac{|V|}{18\sqrt{k}} \geq \frac{|E|/(4.108\sqrt{k})}{18\sqrt{k}} \geq \frac{|E|}{90k}$. Then, for $d_\epsilon = \lceil 180k/\epsilon \rceil$

$$(6)$$

and $l_\epsilon \in \operatorname{argmin}_{l \in \{0,\ldots,d_\epsilon - 1\}} |E[G] \cap E_{d,l}|$, we derive that $\rho(G, H^*_{d_\epsilon, l_\epsilon}) \leq (1 + 180k/d_\epsilon)\rho(G, G^*) \leq (1 + \epsilon)\rho(G, G^*)$.

It remains to prove that $H^*_{d_\epsilon, l_\epsilon}$ can be found efficiently. Since the diameter of the graph $H_{d_\epsilon, l_\epsilon}$ is bounded by d_ϵ, the treewidth is bounded by $6(d_\epsilon + 1)(k + 1)$ and, therefore, we can construct a nice tree decomposition of treewidth at most $5 \cdot (6(d_\epsilon + 1)(k + 1)) + 4$ in time $2^{O(kd_\epsilon)} n = 2^{O(k^2\epsilon^{-1})} n$ [7]. Having a nice tree decomposition of width at most $6(d + 1)(k + 1)$, by Theorem 1, $H^*_{d_\epsilon, l_\epsilon}$ can be computed in time $O\left(2^{6(d_\epsilon + 1)(k+1)\log(6(d_\epsilon + 1)(k+1))}\right) n = 2^{O(k^2\epsilon^{-1}\log(k^2\epsilon^{-1}))} n$.

References

1. Ageev, A.A., Il'ev, V.P., Kononov, A.V., Talevnin, A.S.: Computational complexity of the graph approximation problem. J. Appl. Ind. Math. **1**(1), 1–8 (2007)
2. Arnborg, S., Lagergren, J., Seese, D.: Easy problems for tree-decomposable graphs. J. Algorithms **12**(2), 308–340 (1991)
3. Baker, B.S.: Approximation algorithms for NP-complete problems on planar graphs. J. ACM **41**(1), 153–180 (1994)
4. Bansal, N., Blum, A., Chawla, S.: Correlation clustering. Mach. Learn. **56**(1–3), 89–113 (2004)
5. Ben-Dor, A., Shamir, R., Yakhini, Z.: Clustering gene expression patterns. J. Comput. Biol. **6**(3/4), 281–297 (1999)
6. Böcker, S., Briesemeister, S., Bui, Q.B.A., Truß, A.: Going weighted: parameterized algorithms for cluster editing. Theor. Comput. Sci. **410**(52), 5467–5480 (2009)
7. Bodlaender, H.L., Drange, P.G., Dregi, M.S., Fomin, F.V., Lokshtanov, D., Pilipczuk, M.: A c^k n 5-approximation algorithm for treewidth. SIAM J. Comput. **45**(2), 317–378 (2016)
8. Charikar, M., Guruswami, V., Wirth, A.: Clustering with qualitative information. J. Comput. Syst. Sci. **71**(3), 360–383 (2005)
9. Cygan, M., Fomin, F.V., Kowalik, L., Lokshtanov, D., Marx, D., Pilipczuk, M., Pilipczuk, M., Saurabh, S.: Parameterized Algorithms. Springer, Heidelberg (2015)
10. Deshmukh, B.S., Mankar, V.H.: Segmentation of microscopic images: a survey. In: 2014 International Conference on Electronic Systems, Signal Processing and Computing Technologies (ICESC), pp. 362–364, January 2014
11. Dujmović, V., Eppstein, D., Wood, D.R.: Genus, treewidth, and local crossing number. In: Di Giacomo, E., Lubiw, A. (eds.) GD 2015. LNCS, vol. 9411, pp. 87–98. Springer, Heidelberg (2015). doi:10.1007/978-3-319-27261-0_8
12. Dujmovic, V., Morin, P., Wood, D.R.: Layered separators in minor-closed families with applications. CoRR, abs/1306.1595 (2013)
13. Eppstein, D.: Diameter and treewidth in minor-closed graph families. Algorithmica **27**(3), 275–291 (2000)
14. Idrees, H., Saleemi, I., Seibert, C., Shah, M.: Multi-source multi-scale counting in extremely dense crowd images. In: 2013 IEEE Conference on Computer Vision and Pattern Recognition, 23–28 June 2013, Portland, OR, USA, pp. 2547–2554 (2013)
15. Il'ev, V.P., Il'eva, S.D., Navrotskaya, A.A.: Approximation algorithms for graph approximation problems. J. Appl. Ind. Math. **5**(4), 569–581 (2011)
16. Snell, J.L., Kemeny, J.G.: Mathematical Models in the Social Sciences. Introduction to Higher Mathematics. Ginn, Boston (1962)
17. Kim, S., Yoo, C.D., Nowozin, S., Kohli, P.: Image segmentation usinghigher-order correlation clustering. IEEE Trans. Pattern Anal. Mach. Intell. **36**(9), 1761–1774 (2014)
18. Klein, P.N., Mathieu, C., Zhou, H.: Correlation clustering and two-edge-connected augmentation for planar graphs. In: 32nd International Symposium on Theoretical Aspects of Computer Science, STACS 2015, 4–7 2015, Garching, Germany, pp. 554–567, March 2015
19. Kloks, T.: Treewidth, Computations and Approximations. LNCS, vol. 842. Springer, Heidelberg (1994)
20. Pach, J., Tóth, G.: Graphs drawn with few crossings per edge. Combinatorica **17**(3), 427–439 (1997)

21. Rose, D.J.: On simple characterizations of k-trees. Discrete Math. **7**(3–4), 317–322 (1974)
22. Yarkony, J., Ihler, A., Fowlkes, C.C.: Fast planar correlation clustering for image segmentation. In: Fitzgibbon, A., Lazebnik, S., Perona, P., Sato, Y., Schmid, C. (eds.) ECCV 2012. LNCS, vol. 7577, pp. 568–581. Springer, Heidelberg (2012). doi:10.1007/978-3-642-33783-3_41
23. Zahn, C.T.: Approximation symmetric relations by equivalence relations. J. Soc. Ind. Appl. Math. **12**(4), 840–847 (1964)

Bin Packing with Colocations

Jean-Claude Bermond[1], Nathann Cohen[2], David Coudert[1],
Dimitrios Letsios[1(✉)], Ioannis Milis[3], Stéphane Pérennes[1],
and Vassilis Zissimopoulos[4]

[1] Université Côte d'Azur, Inria, CNRS, I3S, Sophia Antipolis, France
dletsios@unice.fr
[2] CNRS and University of Paris Sud, Orsay, France
[3] Department of Informatics, Athens University of Economics and Business,
Athens, Greece
[4] Department of Informatics and Telecommunications,
National and Kapodistrian University of Athens, Athens, Greece

Abstract. Motivated by an assignment problem arising in MapReduce
computations, we investigate a generalization of the Bin Packing problem which we call *Bin Packing with Colocations Problem*. We are given
a weigthed graph $G = (V, E)$, where V represents the set of items with
positive integer weights and E the set of related (to be colocated) items,
and an integer q. The goal is to pack the items into a minimum number
of bins so that (i) for each bin, the total weight of the items packed in
this bin is at most q, and (ii) for each edge $(i, j) \in E$ there is at least
one bin containing both items i and j.

We first point out that, when the graph is unweighted (i.e., all the
items have equal weights), the problem is equivalent to the q-clique problem, and when furthermore the graph is a clique, optimal solutions are
obtained from Covering Designs. We prove that the problem is strongly
NP-hard even for paths and unweighted trees. Then, we propose approximation algorithms for particular families of graphs, including: a $(3+\sqrt{5})$-
approximation algorithm for complete graphs (improving a previous ratio
of 8), a 2-approximation algorithm for paths, a 5-approximation algorithm for trees, and an $(1 + O(\log q/q))$-approximation algorithm for
unweighted trees. For general graphs, we propose a $3 + 2\lceil mad(G)/2\rceil$-
approximation algorithm, where $mad(G)$ is the maximum average degree
of G. Finally, we show how to convert any approximation algorithm for
Bin Packing (resp. Densest q-Subgraph) problem into an approximation
algorithm for the problem on weighted (resp. unweighted) general graphs.

This work is partially supported by ANR project Stint under reference ANR-13-
BS02-0007, ANR program "Investments for the Future" under reference ANR-11-
LABX-0031-01, the Research Center of Athens University of Economics and Business
(RC-AUEB), and the Special Account for Research Grants of National and Kapodistrian University of Athens.

© Springer International Publishing AG 2017
K. Jansen and M. Mastrolilli (Eds.): WAOA 2016, LNCS 10138, pp. 40–51, 2017.
DOI: 10.1007/978-3-319-51741-4_4

1   Introduction

In this paper, we study the following generalization of the classical *Bin Packing* problem, which we call *Bin Packing with Colocations Problem* (BPCP). We are given a weigthed graph $G = (V, E)$, where $V = \{1, 2, \ldots, n\}$ represents the set of items with positive integer weights $w_1, w_2, \ldots, w_n$ and E the set of related (to be colocated) items, and an integer capacity q for bins. The goal is to pack the items into a minimum number of bins so that (i) for each bin the total weight of the items packed in it is at most q, and (ii) for each edge $(i, j) \in E$ there is at least one bin containing both items i and j. Due to the last constraint of colocating pairwise related items, we assume that, for each edge $(i, j) \in E$, $w_i + w_j \leq q$, for otherwise our problem has no feasible solution. Note also that in a feasible solution (copies of) a vertex (item) might be packed into more than one bin.

Our initial motivation for studying BPCP was the work of Afrati et al. [1,2] on an assignment problem in MapReduce computations. In such computations, the outputs of the mappers, of the form $\langle key - value \rangle$, are assigned to the reducers and each reducer applies a reduce function to a single *key* and its associated list of *value*'s to produce its output. However, a reducer (in fact, the machine executing it) is subject to capacity constraints (e.g. memory size), which limits the total size of data assigned to it. Moreover, for each required output, there must be a reducer receiving all inputs necessary to compute its output. For a family of problems arising in this context, an output depends on pairwise related inputs, i.e., a situation captured by the colocation constraint in BPCP.

More generally, the BPCP models any practical situation where context-related entities of given sizes must be assigned to physical resources of limited capacity while fulfilling pairwise colocation constraints. For instance, when computer files are placed into memory blocks of fixed size, it is natural to ask for the colocation of pairwise related files (for example, sharing a common attribute) in the same memory block. Moreover, in large data centers, file colocation is essential for data chunks which are highly likely to be accessed together.

BPCP is clearly a generalization of the Bin Packing problem, which is the particular case $E = \emptyset$. As another example of this relation, consider BPCP on a star graph with $n + 1$ vertices (items), where the central vertex has weight w_0 and the bin capacity is $q + w_0$. Obviously, BPCP is equivalent to the Bin Packing problem with input the n leaves items (with their weights) and bin capacity q. In contrast to the Bin Packing problem, BPCP remains interesting even when all the items have the same weight and we refer to this case as *Unweighted BPCP* (U-BPCP). It is easy to see that U-BPCP is trivial on a star graph or on a path, but we will prove that it becomes $\mathcal{NP}$-hard even for trees.

Interestingly, U-BPCP for complete graphs falls in the well known area of Combinatorial Design theory (the interested reader is referred to [7] for a survey of this area). In this context, given a set V of n elements, a 2-$(n, q, 1)$-covering design (see [10,12]) abbreviated here as (n, q)-covering is a collection of subsets, which are called *blocks*, such that each block has q elements and every pair of distinct elements of V has to appear together in *at least* one block. An (n, q)-

covering is nothing else than a solution to U-BPCP for complete graphs. In the case of perfect coverings, where each pair appears in exactly one block, the (n, q)-covering is called a BIBD$(n, q, 1)$ or a 2-$(n, q, 1)$ design and a lot of work has been done on necessary and sufficient conditions for the existence of such designs (see [7]). The main observation here is that, if a 2-$(n, q, 1)$-design exists, then it is an optimal solution to U-BPCP for complete graphs.

Furthermore, BPCP generalizes the so called q-Clique Covering Problem studied by Goldschmidt et al. [8]. In their context, a q-clique of a graph G is an induced subgraph with at most q vertices. The objective is to find the minimum number of such q-cliques, such that every edge and every vertex of G is included in at least one q-clique. This corresponds exactly to U-BPCP.

Related Work. Afrati et al. [1,2] studied BPCP for complete and complete bipartite graphs. For both cases, they proved that BPCP is $\mathcal{NP}$-hard, via a reduction from the Partition problem, and they proposed greedy approximation algorithms with ratio 8. For the U-BPCP, they also proposed a $(2 + \epsilon)$-approximation algorithm in the case of complete graphs.

Goldschmidt et al. [8] have proposed approximation algorithms for the q-Clique Covering Problem which corresponds to U-BPCP on general graphs. In fact, for the special cases where $q = 3$ and $q = 4$ (q is the bin capacity), they obtained approximation ratios $7/5$ and $7/3$, respectively. When the bin capacity is arbitrary, they showed that the problem admits an $O(q)$-approximation algorithm.

As described above, U-BPCP on complete graphs is equivalent to finding an (n, q)-covering with the minimum number of blocks (bins). Therefore, the results obtained in combinatorial design theory apply to U-BPCP on complete graphs too and we elaborate on them in Sect. 2.

Finally, as BPCP is a generalization of the Bin Packing problem, we refer the reader to [6] for a recent review of the latter problem. Bin Packing is APX-hard as it is NP-hard to decide between cost 2 and cost 3 (Partition). Simple greedy algorithms as Next-Fit, First-Fit and First-Fit Decreasing achieve approximation ratios of 2, 1.7 and 1.5, respectively. Moreover, it admits asymptotic polynomial-time approximation schemes (APTAS).

Contributions. In Sect. 2, following the work of Afrati et al. [1,2], we begin with the study of U-BPCP and BPCP on complete graphs. We start with U-BPCP where we can use the results obtained on covering. We first present an algorithm similar to the one presented in [1,2] for the case q even, but our analysis is tighter. Our algorithm achieves an approximation ratio less than 2 when q is even and $n \geq q^2/2$. This algorithm can be generalized and, by using $(n, 3)$-coverings (resp. $(n, 4)$-coverings) we get an approximation ratio less than $3/2$ (resp. $5/4$) when q is multiple of 3 (resp. multiple of 4) and $n \geq q^2$. For BPCP an 8-approximation algorithm was given in [1,2]; we propose a new approximation algorithm with ratio 6 and a refined one with ratio $(3 + \sqrt{5})$.

Thereafter, we move our attention to other interesting types of graphs. In Sect. 3, we show that BPCP is strongly $\mathcal{NP}$-hard even on paths and we propose a 2-approximation algorithm for this case. In Sect. 4, we show that U-BPCP is $\mathcal{NP}$-hard on trees and we propose an algorithm which asymptotically achieves an approximation ratio of $(1 + \epsilon)$, where $\epsilon = O(\log q/q)$. Moreover, we propose a greedy 5-approximation algorithm for BPCP on trees. In Sect. 5, we study U-BPCP and BPCP on general graphs. Extending our ideas for BPCP on trees to BPCP on general graphs we derive an approximation algorithm with ratio $3 + 2\lceil mad(G)/2 \rceil$, where $mad(G)$ is the maximum average degree of the graph. This algorithm is efficient for sparse graphs and, for example, it achieves a 9-approximation ratio for BPCP on planar graphs. Then, based on a simple greedy approach, and given any ρ-approximation algorithm for the Bin Packing problem, we obtain a $\rho \cdot \Delta$-approximation algorithm for BPCP on general graphs, where Δ is the maximum degree of the graph. Finally, we show that any ρ-approximation for Densest q-Subgraph problem can be converted to a $\rho \cdot \log n$-approximation algorithm for the U-BPCP on general graphs.

Due to space limitations, all proofs are deferred to the research report version of the paper [4].

2   Complete Graphs

In the following we observe that U-BPCP on complete graphs is closely related to the theory of combinatorial designs (see [7]). For this reason, we briefly survey some fundamental results known in this area.

Given a set V of n elements, a 2-$(n, q, 1)$-*design* or $BIBD(n, q, 1)$ is a collection of subsets of V, called *blocks*, such that every pair of distinct elements appears together in exactly one block. In other words it corresponds to a partition of the edges of K_n into K_q. In such a design, every element appears in $(n-1)/(q-1)$ blocks and the number of blocks must be equal to $n(n-1)/q(q-1)$. Since these numbers must be integers, two necessary conditions for the existence of a 2-$(n, q, 1)$-design are $(n - 1) \equiv 0 \mod q - 1$ and $n(n - 1) \equiv 0 \mod q(q - 1)$. These necessary conditions have been proved to be sufficient for certain values of n and q (see [7]), for instance when $q = 3$ (known as Steiner triple systems) and $q = 4, 5$ or when q is a power of a prime and $n = q^2$ or $n = q^2 + q + 1$. Furthermore, Wilson [13] has proved that these necessary conditions are also sufficient when n is large enough. Still, in many cases these conditions do not guarantee the existence of a 2-$(n, q, 1)$-design; for example, as guessed by Euler both a 2-$(36, 6, 1)$-design or a 2-$(43, 7, 1)$-design do not exist [7].

Clearly, a 2-$(n, q, 1)$-design is an optimal solution for U-BPCP on a complete graph with n vertices and bin capacity q. Note that this relation was not observed by Afrati et al. [1,2] who rediscovered basic results of design theory such as the existence of some $(n, 3, 1)$-design and the existence of projective planes.

The notion of 2-$(n, q, 1)$-design has been also extended to packing and covering designs (see the survey [10] or chapter IV.8 in Handbook of Designs [12]). Given a set V of n elements, a 2-$(n, q, 1)$-covering design (see [10,12]) abbreviated here as (n, q)-covering is a collection of subsets, which are called *blocks*, such that each block has q elements and every pair of distinct elements of V

appears together in *at least* one block. An (n, q)-covering is nothing else than a solution to U-BPCP for complete graphs.

In the literature, there exists a significant amount of work on computing the the minimum number of blocks in an (n, q)-covering, called the covering number and denoted $C(n, q)$. Therefore, for U-BPCP on complete graphs, the number of bins of an optimal solution is equal to $C(n, q)$.

In what follows, let $L(n, q) = \left\lceil \frac{n}{q} \left\lceil \frac{n-1}{q-1} \right\rceil \right\rceil$; this quantity will serve for lower bounding the number of used bins in an optimal solution.

Lemma 1. *It holds that $C(n, q) \geq L(n, q)$. Furthermore, if $(n - 1) \equiv 0 \mod (q - 1)$ and $n(n - 1) \equiv 1 \mod q$, then $C(n, q) \geq L(n, q) + 1$.*

The exact values of $C(n, q)$ have been determined only in some cases (see [10,12]). For example, the exact value of $C(n, q)$ is known for $n <= 3q$ and for $q = 2, 3, 4$ where we have:

- $C(n, 2) = L(n, 2) = \frac{n(n-1)}{2}$ (trivial as a block contains one pair),
- $C(n, 3) = L(n, 3) = \left\lceil \frac{n}{3} \left\lceil \frac{n-1}{2} \right\rceil \right\rceil$, and
- $C(n, 4) = L(n, 4) + \epsilon$, where $\epsilon = 1$ when $n = 7, 9, 10$, $\epsilon = 2$ in the case where $n = 19$, and $\epsilon = 0$, otherwise.

Finally, the following theorem, which has been proved by Rödl [11] via probabilistic methods, bounds $C(n, q)$ asymptotically. Interestingly, it answered a conjecture of Erdös and Hanani (see Chap. 4 of [3] for a proof).

Theorem 1 (Rödl [11]). *For any fixed q, it holds that $C(n, q) \leq (1 + o(1))L(n, q)$, where the term $o(1)$ approaches zero as n tends to infinity.*

Unfortunately, this theoretical result does not give answers for practical values of n and q and, for such cases, we propose some simple greedy algorithms.

2.1   Unweighted case

The main idea for designing an approximation algorithm consists in partitioning the items into $g = \lceil n/\lfloor q/k \rfloor \rceil$ groups of equal size $\lfloor q/k \rfloor$ (except possibly one), where k is a chosen positive integer for which aand to use a (g, k)-covering. All the items of such a group are then considered as one element and we cover the pairs of groups with blocks of size k. For each block, we use a bin consisting of all items of the groups in the block. As a block contains k groups, a bin will contain at most $k \lfloor q/k \rfloor \leq q$ items. Furthermore, each pair of items belongs to some bin. Indeed, consider a pair $\{i, j\}$; i belongs to some group A and j to some group B. Then the pair $\{i, j\}$ belongs to the bin associated to the block containing the pair of groups A and B if A and B are distinct, or to every bin containing A if $A = B$.

The analysis of this general algorithm might be difficult as we have various floors and ceils and also it assumes the existence of a good (g, k)-covering. Moreover, the approximation ratio obtained will depend of the size of the groups;

indeed the pairs of items belonging to the same group will be repeated many times. So we have interest to choose a large k, but very few (g, k)-coverings are known for large k.

For $k = 2$, a case for which a trivial $(g, 2)$-covering exists, we get Algorithm 1. This is similar with the one of [1,2] for even values of q and simpler than their algorithm for odd values of q. However, here we present a tighter analysis resulting in slightly better approximation ratios.

Algorithm 1 (U-BPCP, complete graphs)

1: Partition the items into g groups each of size $\lfloor q/2 \rfloor$, but at most one group.
2: Pack every pair of groups into a bin.

Theorem 2. *Algorithm 1 achieves approximation ratios of* $2\frac{q-1}{q} + \frac{q-2}{n}$, *if q is even, and* $2\frac{q}{q-1} + \frac{q-1}{n}$, *if q is odd, for the U-BPCP on complete graphs.*

Note that, by Theorem 2, we have an approximation ratio less than 2, when q is even and $n \geq q^2/2$. When q is odd, the algorithm has no interest for $q = 3$ and $n \leq 3q$ as we know in that case the exact value of $C(n, q)$. So, we will use the algorithm only for $q \geq 5$ and $n > 3q$, in which case the approximation ratio is less than $17/6$. Note also that when q is large and n tends to infinity the ratio is near to 2.

We can also analyze the general algorithm described above for $k = 3$ (resp. $k = 4$) and q is a multiple of 3 (resp. 4), to get an approximation ratio at most $3/2$ (resp. $5/4$). More generally, for any k, if n is a multiple of q and there exists a (g, k)-covering, for $g = \lceil \frac{kn}{q} \rceil$, we get a $\frac{k}{k-1}$-approximation ratio.

2.2 Weighted Case

In this section, we extend the previous ideas to the BPCP on complete graphs by using an appropriate grouping of jobs. Initially, we present a 6-approximation algorithm via a simple grouping which we then improve via a more enhanced grouping. Our analysis uses the lower bound on the optimal number of bins, b^*, provided by the next lemma.

Lemma 2. *For the BPCP on a complete graph it holds that*
$b^* \geq \frac{1}{q} \sum_{i=1}^{n} w_i \lceil \frac{s-w_i}{q-w_i} \rceil > \frac{s^2}{q^2}$, *where* $s = \sum_{i=1}^{n} w_i$.

In [1,2], the authors showed that Algorithm 1 can also be used for weighted graph and gives an approximation ratio of 8. In Algorithm 2, we use a better grouping which achieves a feasible solution and improves the approximation ratio to 6. Note that, in Algorithm 2, we suppose w.l.o.g. that all the weights are at most $q/2$. Indeed, there can be at most one item of weight greater than $q/2$ as the input graph is complete. In such a case, the large item can be packed independently with all the other items and the remaining pairs of items can be packed with Algorithm 2.

Algorithm 2 (BPCP, complete graphs)

1: Partition the items into three types of groups A, B, C;
   the size of a group $s_{A_i}, s_{B_i}, s_{C_i}$ is the sum of the weights of the items in the group:

   - α groups of type A; $A_i \in A$ has size $\frac{q}{3} < s_{A_i} \leq \frac{q}{2}$
   - β groups of type B; $B_i \in B$ has size $\frac{q}{4} < s_{B_i} \leq \frac{q}{3}$
   - γ groups of type C; $C_i \in C$ has size $s_{C_i} \leq \frac{q}{4}$

2: Form bins containing either a pair of groups or when possible three groups.

Theorem 3. *Algorithm 2 achieves an approximation ratio of 6 for the BPCP on complete graphs.*

We can refine the above idea, by partitioning the items into four types of groups, to get an even better approximation ratio.

Theorem 4. *There is a $(3 + \sqrt{5})$-approximation algorithm for the BPCP on complete graphs.*

3   Paths

In this section we consider the BPCP on paths; recall that U-BPCP is trivial on paths. We first show that the BPCP on paths is strongly $\mathcal{NP}$-hard via a reduction from the Bin Packing problem.

Theorem 5. *The BPCP on paths is strongly $\mathcal{NP}$-hard.*

We also present a 2-approximation algorithm by a reduction to the shortest path problem on an appropriate directed graph and the use of the Next-Fit algorithm for the Bin Packing problem. Starting from a path $G = (V, E)$, with $V = \{1, 2, \ldots, n\}$ and $E = \{(i, i+1) | 1 \leq i \leq n-1\}$, we construct an auxiliary weighted directed graph $\overrightarrow{G}$ which contains a node for each vertex $i \in V$. Then, for each pair (i, j) such that $1 \leq i < j \leq n$, we denote by $W(i, j) = \sum_{k=i}^{j} w_k$ the total weight of the vertices $i, i+1, \ldots, j$ and, if $W(i, j) \leq q$, then $\overrightarrow{G}$ contains an arc (i, j) of weight $W(i, j)$. Clearly, any $(1, n)$-path (i.e. a path from node 1 to node n) P of $\overrightarrow{G}$ corresponds to a feasible solution of our problem; for each arc $(i, j) \in P$ we use a bin to pack vertices $i, i+1, \ldots, j$. For a path P of $\overrightarrow{G}$, we denote by $W(P) = \sum_{(i,j) \in P} W(i, j)$ its total weight. The following lemma provides a lower bound on the optimal number of bins, b^*, for the BPCP on paths which we use in our analysis.

Lemma 3. *For the BPCP on paths it holds that $b^* \geq \frac{1}{q} \cdot W(P^*)$, where P^* is a minimum weight $(1, n)$-path in the auxiliary graph $\overrightarrow{G}$.*

Our Algorithm 3 considers each arc in a minimum weight $(1, n)$-path in $\overrightarrow{G}$ as an item for the Bin Packing problem and packs them using the Next-Fit algorithm.

Algorithm 3 (BPCP, paths)

1: Find a minimum weight $(1, n)$-path P^* in $\overrightarrow{G}$.
2: For each arc $(i, j) \in P^*$, create an item of weight $W(i, j)$.
3: Pack the new items using the Next-Fit algorithm.

Using Lemma 3 and the fact that the Next-Fit algorithm packs a set of items of total weight W into at most $2\lceil W/q \rceil$ bins of capacity q we get the next theorem.

Theorem 6. *Algorithm* 3 *achieves an approximation ratio of 2 for the BPCP on paths.*

4   Trees

In this section we deal with both U-BPCP and BPCP on trees. We show that U-BPCP is $\mathcal{NP}$-hard and that it admits an $(1 + \epsilon)$-approximation algorithm. We also propose a greedy 5-approximation algorithm for BPCP on trees.

4.1   Unweighted Case

We show first that U-BPCP on trees is $\mathcal{NP}$-hard via a reduction from the 3-Partition problem which is known to be $\mathcal{NP}$-hard even for polynomially bounded parameters.

Theorem 7. *The U-BPCP on trees is $\mathcal{NP}$-hard.*

For our approximation algorithm, let G be the input tree of our problem and suppose that the edges of G are oriented away from some arbitrary node which is picked as the root and we obtain a directed tree T. A key ingredient for the description of our algorithm is the notion of an *eligible subtree*. Given a directed tree $T = (V(T), E(T))$, an eligible subtree T' is a subtree of T rooted at some vertex $i \in V(T)$ such that, the forest $\overline{T} = ((V(T) \setminus V(T')) \cup \{i\}, E(T) \setminus E(T'))$ consists of a single tree. That is, the removal of all the edges and all the vertices of T', but i, leaves $\overline{T}$ connected. We define the *size* of a tree T as the number of vertices that it contains and we denote it by $s(T)$. The following decomposition lemma is critical for designing our algorithm.

Lemma 4. *There exists an eligible subtree T' of a tree T of size $k/2 \leq s(T') \leq k$, for each $k \in [1, s(T)]$.*

We assume, for convenience, that the bin capacity q is a power of 2, i.e. $q = 2^a$ for some integer $a > 0$, but our analysis can be extended to arbitrary values of q. We also denote by b the number of bins that our algorithm uses. The algorithm starts with the initial tree G and, gradually, it packs vertices into bins and removes vertices whose incident edges have been covered until a feasible solution is produced. More specifically, it consists of b phases and, in each phase, a steps are performed. In the k-th phase, $1 \leq k \leq b$, the algorithm computes the content of the k-th bin, say B_k. During the algorithm's execution, we denote by f_k the free space of bin B_k and by T the current remaining tree (whose edges have not been packed before). In the beginning of the i-th step of the k-th phase, it must be the case that $f_k \leq q/2^{i-1}$. Then, if it also holds that $f_k \geq q/2^i$, based on Lemma 4, the algorithm computes an eligible subtree T' of T (the remaining part of the initial tree) with size $s(T') \in [q/2^{i+1}, q/2^i]$ and it packs T' in B_k. Moreover, the vertices of T', apart from the root, as well as the edges are removed from T. If there is sufficient space, then a second eligible subtree of the same bounded dimension is also computed, is packed in B_k and is removed from T. In this way, at the end of the i-th step, it holds that $f_k \leq q/2^i$ (Lemma 5). The algorithm proceeds until T becomes the empty graph.

Algorithm 4 (U-BPCP, trees, $q = 2^a$)

1: T: directed tree obtained by orienting the edges of G
2: $k = 1$, $f_k = q$
3: **while** $E(T) \neq \emptyset$ **do**
4:     **for** $i = 1, 2, \ldots, a$ **do**
5:         Repeat twice:
6:         **if** $f_k \geq q/2^i$ **then**
7:             Compute an eligible subtree T' such that $s(T') \in [q/2^{i+1}, q/2^i]$.
8:             Pack $V(T')$ in bin B_k and remove T' from T.
9:     $k = k + 1$
10: Return the solution found.

Lemma 5. *At the end of the i-th step in the k-th phase, it holds that $f_k \leq q/2^i$, for $1 \leq i \leq a$ and $1 \leq k \leq b$.*

Theorem 8. *Algorithm 4 achieves asymptotically an approximation ratio of $(1 + \epsilon)$, where ϵ is $O(\log q/q)$, for U-BPCP on trees.*

4.2   Weighted Case

In what follows, we present a greedy 5-approximation algorithm for BPCP on trees. We consider a tree $G = (V, E)$ and we assume again that the edges are oriented away from some node $r \in V$ which is chosen arbitrarily as the root and we obtain a directed tree T. The algorithm produces a feasible solution by considering T. Initially, every node $i \in V$ is packed independently together with all its children so as to ensure feasibility of the obtained solution. More specifically, for each $i \in V$, all vertices in its out-neighborhood $\Gamma^+(i)$ are packed into bins of capacity $q - w_i$ according to the First-Fit Decreasing algorithm. Then, the content of every such bin together with vertex i is considered as one item for the Bin Packing problem and they are packed using the Next-Fit algorithm.

Algorithm 5 (BPCP, trees)

1: **for** each $i \in V$ **do**
2:     Pack the vertices in $\Gamma^+(i)$ into bins of capacity $q - w_i$ using the First-Fit Decreasing algorithm.
3:     For each bin containing a subset S of items, create an item of size $\sum_{i \in S} w_i$.
4: Pack the created items using the Next-Fit algorithm.

It is known that number of bins used by First-Fit Decreasing algorithm for the Bin Packing problem is at most $3/2$ times the number of the optimal number of bins. Using this fact, we can bound the number of copies of each vertex when it is packed with its children and we get the next theorem.

Theorem 9. *Algorithm* 5 *achieves an approximation ratio of* 5 *for BPCP on trees.*

5   General Graphs

In this section we deal with the BPCP and U-BPCP on a general graph $G = (V, E)$. We first deal with BPCP and we present two approximation algorithms. The first one extents our approach for the BPCP on trees to general graphs and gives an approximation ratio which is efficient for the BPCP on sparse graphs. The second one considers each edge $(i, j) \in E$ as an item of the Bin Packing problem of weight $w_i + w_j$ and gives an approximation ratio of $O(\Delta)$, where Δ is the maximum degree of the graph. Then, we move to the U-BPCP and we present an approximation algorithm based on its relation with Densest q-Subgraph problem.

5.1   Weighted Case

We first, extend our approach for the BPCP on trees to BPCP on a general graph $G = (V, E)$. More specifically, we construct an orientation D of the graph G and for each vertex $i \in V$ we consider its in- and out-neighborhood in D. Recall that in BPCP on trees each node is packed with its children and in one more bin with its parent. In the BPCP on general graphs, each node is packed with the vertices in its out-neighborhood and with each one of the vertices in its in-neighborhood in different bins. Using similar arguments as in the proof of Theorem 9 we obtain an approximation ratio of $3 + 2\Delta^-(D)$ where $\Delta^-(D)$ is the maximum in-degree of D.

The maximum average degree $mad(G)$ of the input graph G is the maximum of the average degrees $ad(H) = 2|E(H)|/|V(H)|$ taken over all subgraphs H of G, i.e., $mad(G) = \max_{H \subseteq G} \left\{ \frac{2|E(H)|}{|V(H)|} \right\}$. By applying the approach of Hakimi [9], we can construct, in polynomial time, an orientation D of a general undirected graph G, with maximum in-degree $\Delta^-(D) \leq \lceil mad(G)/2 \rceil$. Using this result we get the next theorem.

Theorem 10. *There is a $3 + 2\lceil mad(G)/2 \rceil$-approximation algorithm for the BPCP on general graphs.*

In the case of planar graphs, it holds that $mad(G) < 6$ and we obtain a 9-approximation. More generally, any class of H-minor-free graphs have bounded maximum average degree.

Next, we present an approximation algorithm for BPCP on a general graph $G = (V, E)$, which uses a ρ-approximation algorithm $\mathcal{A}$ for the Bin Packing problem. We denote by Δ the maximum degree of G.

Initially, we obtain a lower bound by packing the edges of the input graph $G = (V, E)$ instead of its vertices. Specifically, for each edge $(i, j) \in E$, we create an item $I_{i,j}$ of weight $w_i + w_j$. Let I be the set of all such items and consider the instance (I, q) of the Bin Packing problem. Clearly, any feasible packing of (I, q) is a feasible solution for BPCP in general graphs. So, we get the following lemma.

Lemma 6. *Let b^* and b_e^* be the optimal numbers bins for BPCP and the Bin Packing problem (I, q), respectively. Then, it holds that $b_e^* \leq \Delta \cdot b^*$.*

Algorithm 6 (BPCP, general graphs)

1: For each edge $(i, j) \in E$, create an item of weight $w_i + w_j$.
2: Pack the items with $\mathcal{A}$ into bins of capacity q.

Then, Lemma 6 implies the next theorem.

Theorem 11. *Algorithm 6 achieves an approximation ratio of $\rho \cdot \Delta$ for BPCP on general graphs, given a ρ-approximation algorithm for the Bin Packing problem.*

Recall that the Bin Packing problem admits several greedy constant-factor approximation algorithms as well as an APTAS (Asymptotic Polynomial-Time Approximation Scheme).

5.2   Unweighted Case

In what follows, we present an approximation algorithm for U-BPCP on general graphs by using a ρ-approximation algorithm $\mathcal{A}$ for the Densest q-Subgraph problem (i.e. finding a set of q vertices with the maximum number of edges in the subgraph induced by them). More specifically, the algorithm packs repeatedly densest q-subgraphs of G and removes the covered edges. The procedure goes on until all edges are covered, as in Algorithm 7 below.

Theorem 12. *Algorithm 7 is $\rho \cdot \log n$-approximate for U-BPCP on general graphs, given a ρ-approximation algorithm for Densest q-Subgraph problem.*

The best known approximation algorithm for the Densest q-Subgraph problem was proposed by [5] and its approximation ratio is $O(n^{1/4})$. Therefore, Theorem 12 implies a $O(n^{1/4} \cdot \ln n)$-approximation algorithm for U-BPCP.

Algorithm 7 (U-BPCP, general graphs)

1: **while** $E \neq \emptyset$ **do**
2: Run $\mathcal{A}$ and let $D = (V', E')$, $|V'| = q$ the resulting densest q-subgraph.
3: Pack the vertices of V' into a new bin.
4: $G = (V, E \setminus E')$.

References

1. Afrati, F.N., Dolev, S., Korach, E., Sharma, S., Ullman, J.D.: Assignment of different-sized inputs in mapreduce. In: Proceedings of the EDBT/ICDT Conference, Brussels, Belgium, pp. 28–37 (2015)
2. Afrati, F.N., Dolev, S., Korach, E., Sharma, S., Ullman, J.D.: Assignment problems of different-sized inputs in mapreduce. CoRR, abs/1507.04461 (2015)
3. Alon, N., Spencer, J.H.: The Probabilistic Method, 4th edn. Wiley, New York (2016)
4. Bermond, J.-C., Cohen, N., Coudert, D., Letsios, D., Milis, I., Pérennes, S., Zissimopoulos, V.: Bin packing with colocations. Technical report hal-01381333, Inria, I3S (2016). https://hal.inria.fr/hal-01381333
5. Bhaskara, A., Charikar, M., Chlamtac, E., Feige, U., Vijayaraghavan, A.: Detecting high log-densities: an $O(n^{1/4})$ approximation for densest k-subgraph. In: ACM Symposium on Theory of Computing (STOC), pp. 201–210 (2010)
6. Coffman Jr., E.G., Csirik, J., Galambos, G., Martello, S., Vigo, D.: Bin packing approximation algorithms: survey and classification. In: Handbook of Combinatorial Optimization, pp. 455–531 (2013)
7. Colbourn, C., Dinitz, J. (eds.): Handbook of Combinatorial Designs: Discrete Mathematics and its Applications, vol. 42, 2nd edn. Chapman and Hall/CRC, Boca Raton (2006)
8. Goldschmidt, O., Hochbaum, D.S., Hurkens, C.A.J., Yu, G.: Approximation algorithms for the k-clique covering problem. SIAM J. Discrete Math. **9**(3), 492–509 (1996)
9. Hakimi, S.L.: On the degrees of the vertices of a directed graph. J. Frankl. Inst. **279**, 290–308 (1965)
10. Mills, W., Mullin, R.: Coverings and packings. In: Colbourn, C., Dinitz, J. (eds.) Contemporary Design Theory: A Collection of Surveys, pp. 371–399. Wiley, New York (1992)
11. Rödl, V.: On a packing and covering problem. Eur. J. Comb. **6**, 69–78 (1985)
12. Stinson, D.R.: Coverings. In: Colbourn, C., Dinitz, J. (eds.) The CRC Handbook of Combinatorial Designs. Discrete Mathematics and Its Applications, vol. 42, 2nd edn, pp. 260–265. CRC Press, Boca Raton (2006). chapter IV.8
13. Wilson, R.M.: Decomposition of complete graphs into subgraphs isomorphic to a given graph. Congres. Numer. **15**, 647–659 (1976)

Batch Coloring of Graphs

Joan Boyar[1], Leah Epstein[2], Lene M. Favrholdt[1], Kim S. Larsen[1(✉)], and Asaf Levin[3]

[1] Department of Mathematics and Computer Science,
University of Southern Denmark, Odense, Denmark
{joan,lenem,kslarsen}@imada.sdu.dk
[2] Department of Mathematics, University of Haifa, Haifa, Israel
lea@math.haifa.ac.il
[3] Faculty of IE&M, The Technion, Haifa, Israel
levinas@ie.technion.ac.il

Abstract. In graph coloring problems, the goal is to assign a positive integer color to each vertex of an input graph such that adjacent vertices do not receive the same color assignment. For classic graph coloring, the goal is to minimize the maximum color used, and for the sum coloring problem, the goal is to minimize the sum of colors assigned to all input vertices. In the offline variant, the entire graph is presented at once, and in online problems, one vertex is presented for coloring at each time, and the only information is the identity of its neighbors among previously known vertices. In batched graph coloring, vertices are presented in k batches, for a fixed integer $k \geq 2$, such that the vertices of a batch are presented as a set, and must be colored before the vertices of the next batch are presented. This last model is an intermediate model, which bridges between the two extreme scenarios of the online and offline models. We provide several results, including a general result for sum coloring and results for the classic graph coloring problem on restricted graph classes: We show tight bounds for any graph class containing trees as a subclass (e.g., forests, bipartite graphs, planar graphs, and perfect graphs), and a surprising result for interval graphs and $k = 2$, where the value of the (strict and asymptotic) competitive ratio depends on whether the graph is presented with its interval representation or not.

1 Introduction

We study three different graph coloring problems in a model where the input is given in *batches*. In this model of computation an adversary reveals the input graph one batch at a time. Each batch is a subset of the vertex set together with its edges to the vertices revealed in the current batch or in previous batches. After a batch is revealed the algorithm is asked to color the vertices of this batch with colors which are positive integers, the coloring must be valid or *proper*, i.e., neighbors are colored using distinct colors, and this coloring cannot be modified later.

J. Boyar, L.M. Favrholdt and K.S. Larsen—Supported in part by the Danish Council for Independent Research, Natural Sciences, grant DFF-1323-00247, and the Villum Foundation, grant VKR023219.

© Springer International Publishing AG 2017
K. Jansen and M. Mastrolilli (Eds.): WAOA 2016, LNCS 10138, pp. 52–64, 2017.
DOI: 10.1007/978-3-319-51741-4_5

The batch scenario is somewhere between online and offline. In an *offline* problem, there is only one batch, while for an *online* problem, the requests arrive one at a time and have to be handled as they arrive without any knowledge of future events, so each request is a separate batch. Many applications might fall between these two extremes of online and offline. For example, a situation where there are two (or more) deadlines, an early one with a lower price and a later one with a higher price can lead to batches.

When considering a combinatorial problem using batches, we assume that the requests arrive grouped into a constant number k of batches. Each batch must be handled without any knowledge of the requests in future batches. As with online problems, we do not consider the execution times of the algorithms used within one batch; the focus is on the performance ratios attainable. Therefore, our goal is to quantify the extent to which the performance of the solution deteriorates due to the lack of information regarding the requests of future batches. We also investigate how much advance knowledge of the number of batches can help.

The quality of the algorithms is evaluated using competitive analysis. Let $A(\sigma)$ denote the cost of the solution returned by algorithm A on request sequence σ, and let $\text{OPT}(\sigma)$ denote the cost of an optimal (offline) solution. Note that for standard coloring problems, $\text{OPT}(G) = \chi(G)$, where $\chi(G)$ is the chromatic number of the graph G. An online coloring algorithm A is ρ-*competitive* if there exists a constant b such that, for all finite request sequences σ, $A(\sigma) \leq \rho \cdot \text{OPT}(\sigma) + b$. The *competitive ratio* of algorithm A is $\inf\{\rho \mid A \text{ is } \rho\text{-competitive}\}$. If the inequality holds with $b = 0$, the algorithm is strictly ρ-*competitive* and the *strict competitive ratio* is $\inf\{\rho \mid A \text{ is strictly } \rho\text{-competitive}\}$.

The First-Fit algorithm for coloring a graph traverses the list of vertices given in an arbitrary order or in the order they are presented, and assigns each vertex the minimal color not assigned to its neighbors that appear before it in the list of vertices.

Other combinatorial problems have been studied previously using batches. The study of bin packing with batches was motivated by the property that all known lower bound instances have the form that items are presented in batches. The case of two batches was first considered in [9], an algorithm for this case was presented in [6], and better lower bounds were found in [2]. A study of the more general case of k batches was done in [7], and recently, a new lower bound on the competitive ratio of bin packing with three batches was presented in [1]. The scheduling problem of minimizing makespan on identical machines where jobs are presented using two batches was considered in [20].

Graph Classes Containing Trees. The first coloring problem we consider using batches is that of coloring graph classes containing trees as a subclass (e.g., forests, bipartite graphs, planar graphs, perfect graphs, and graphs in general), minimizing the number of colors used. Offline, finding a proper coloring of bipartite graphs is elementary and only (at most) two colors are needed. However, there is no online algorithm with a constant competitive ratio, even for trees. Gyárfás and Lehel [10] show that for any online tree coloring algorithm A and any $n \geq 1$, there is a tree on n vertices for which A uses at least $\lfloor \log n \rfloor + 1$ colors.

The lower bound is matched exactly by First-Fit [11], and hence, the optimal competitive ratio on trees is $\Theta(\log n)$. For general graphs, Halldórsson and Szegedy [13] have shown that the competitive ratio is $\Omega(n/\log n)$.

We show that any algorithm for coloring trees in k batches uses at least $2k$ colors in the worst case, even if the number of batches is known in advance. This gives a lower bound of k on the competitive ratio of any algorithm coloring trees in k batches. The lower bound is tight, since (on any graph, not only trees), a k-competitive algorithm can be obtained by coloring each batch optimally with colors not used in previous batches. Thus, for graph classes containing trees as a subclass, k is the optimal competitive ratio.

Coloring Interval Graphs in Two Batches. Next we consider coloring interval graphs in two batches, minimizing the number of colors used. An interval graph is a graph which can be defined as follows: The vertices represent intervals on the real line, and two vertices are adjacent if and only if their intervals overlap (have a nonempty intersection). If the maximum clique size of an interval graph is ω, it can be colored optimally using ω colors by using First-Fit on the interval representation of the graph, with the intervals sorted by nondecreasing left endpoints. For the online version of the problem, Kierstead and Trotter [15] provided an algorithm which uses at most $3\omega - 2$ colors and proved a matching lower bound for any online algorithm.

The algorithm presented in [15] does not depend on the interval representation of the graph, but the lower bound does, so in the online case the optimal competitive ratio is the same for these two representations (see [14,18] for the current best results regarding the strict competitive ratio of First-Fit for coloring interval graphs). In contrast, when there are two batches, there is a difference. We show tight upper and lower bounds of 2 for the case when the interval representation is unknown and 3/2 when it is known, respectively. Our results apply to both the asymptotic and the strict competitive ratio.

Note that when the interval representation of the graph is used, the batches consist of intervals on the real line (it is not necessary to give the edges explicitly).

Sum Coloring. The sum coloring problem (also called chromatic sum) was introduced in [17] (see [16] for a survey of results on this problem). The problem is to give a proper coloring to the vertices of a graph, where the colors are positive integers, so as to minimize the sum of these colors over all vertices (that is, if the coloring is defined by a function C, the objective is to minimize $\sum_{v \in V} C(v)$).

Bar-Noy et al. [3] study the problem, motivated by the following application: Consider a scheduling problem on an infinite capacity batched machine where all jobs have unit processing time, but some jobs cannot be run simultaneously due to conflicts for resources. If the conflicts are given by a graph where the jobs are vertices and an edge exists between two vertices, if the corresponding jobs cannot be executed simultaneously (and thus each batch of jobs corresponds to an independent set of this graph), the value s of the optimal sum coloring of the graph gives the sum of the completion times of all jobs in an optimal schedule. Dividing by the number of jobs gives the average response time. The problem when restricted to interval graphs is also motivated by VLSI routing [19]. The first problem seems more likely to come in batches than the second.

The sum coloring problem is NP-hard for general graphs [17] and cannot be approximated within $n^{1-\varepsilon}$ for any $\varepsilon > 0$ unless ZPP = NP [3]. Interestingly, there is a linear time algorithm for trees, even though there is no constant upper bound on the number of different colors needed for the minimum sum coloring of trees [17]. For online algorithms, there is a lower bound of $\Omega(n/\log^2 n)$ for general graphs with n vertices [12].

We show tight upper and lower bounds of k on the competitive ratio when there are k batches and k is known in advance to the algorithm. The competitive ratio is higher if k is unknown in advance to the algorithm. We do not give a closed form expression for the competitive ratio in this case, but give tight upper and lower bounds on the order of growth of the competitive ratio and the strict competitive ratio. For any nondecreasing function f, with $f(1) \geq 1$, the optimal competitive ratio for k batches is $O(f(k))$ if the series $\sum_{i=1}^{\infty} \frac{1}{f(i)}$ converges, and it is $\Omega(f(k))$ if the series diverges. Thus, for example, it is $O(k \log k (\log \log k)^2)$ and $\Omega(k \log k \log \log k)$.

Restricting to trees, First-Fit is strictly 2-competitive for the online problem. Thus, First-Fit gives a (strict) competitive ratio of 2 regardless of the number of batches. See for example [4] for results on the strict competitive ratio of First-Fit for other graph classes.

Omitted proofs and details appear in the full paper [5].

2   Graph Classes Containing Trees

In this section, we study the problem of coloring trees in k batches. The results hold for any graph class that contains trees as a special case, including bipartite graphs, planar graphs, perfect graphs, and the class of all graphs. If we want the algorithm to be polynomial time, then we are restricted to graph classes where optimal offline coloring is possible in polynomial time (e.g., perfect graphs [8]).

The construction proving the following lemma resembles that of the lower bound of $\Omega(\log n)$ for the competitive ratio for online coloring of trees [10].

Lemma 1. *For any integer $k \geq 1$, any algorithm for k-batch coloring of trees can be forced to use at least $2k$ colors, even if k is known in advance.*

The following lemma holds for any graph, not only trees.

Lemma 2. *There is a strictly k-competitive algorithm for k-batch coloring, even if k is not known in advance.*

Theorem 1 below follows directly from Lemmas 1 and 2.

Theorem 1. *For any graph class containing trees as a special case, the optimal (strict) competitive ratio for k-batch coloring is k, regardless of whether or not k is known in advance.*

3 Interval Coloring in Two Batches

Since not all trees are interval graphs, the lower bound from the previous section does not apply here. For the case of interval graphs we show the surprising result that while coloring in two batches has a tight bound of 2, the problem becomes easier if we assume that the vertices of the graph are revealed together with their interval representation (and this interval representation of vertices of the first batch cannot be modified in the second batch). The standard results for online coloring of interval graphs do not make this distinction: The lower bound is obtained for the (a priori easier) case where the interval representation of a vertex is revealed to the algorithm when the vertex is revealed, while the upper bound holds even if such a representation is not revealed (the online algorithm only computes a maximum clique size containing the new vertex and applies the First-Fit algorithm on a subset of the vertices). Throughout this section, our lower bounds are with respect to the asymptotic competitive ratio while our upper bounds are for the strict competitive ratio, and thus the results are tight for both measures.

Unknown Interval Representation. We start with a study of the case where the algorithm is guarantied that the resulting graph (at the end of every batch) will be an interval graph, but the interval representation of the vertices of the first batch is not revealed to the algorithm (and may depend on the actions of the algorithm). We show that in this case 2 is the best competitive ratio that can be achieved by an online algorithm.

Theorem 2. *For the problem of 2-batch coloring of interval graphs with unknown interval representation, the optimal (strict) competitive ratio is 2.*

Proof. The upper bound follows from Lemma 2. Each of the two induced subgraphs is an interval graph, and it can be colored optimally in polynomial time even if the interval representation is not given.

   Next, we show a matching lower bound. For a given $q \in \mathbb{N}$, let $N_1 = \binom{4q}{q} + 1$ and $N_2 = \binom{4q}{2q} + 1$. In the first batch, the adversary gives $N_1 + N_2$ pairwise nonoverlapping cliques: N_1 cliques of size q and N_2 cliques of size $2q$.

   Assume that an algorithm uses at most $4q$ colors for the first batch. By the pigeon hole principle, there are two cliques of size q that are colored with the same set $\mathcal{C}_1$ of colors. The vertices of these two cliques will correspond to the intervals $[5, 6]$ and $[9, 10]$, respectively. Similarly, there are two cliques of size $2q$ that are colored with the same set $\mathcal{C}_2$ of colors. For one of these cliques, q vertices will correspond to the interval $[0, 1]$ and the remaining q vertices will correspond to the interval $[0, 3]$. If any of these $2q$ vertices are colored with colors from $\mathcal{C}_1$, they will correspond to the interval $[0, 1]$. We let $\mathcal{C}'_2$ denote the set of colors used on the vertices corresponding to the interval $[0, 3]$. Note that $\mathcal{C}_1 \cap \mathcal{C}'_2 = \emptyset$, and hence, $|\mathcal{C}_1 \cup \mathcal{C}'_2| = 2q$. For the other of these two cliques, the q vertices colored with $\mathcal{C}'_2$ will correspond to the interval $[12, 15]$ and the remaining q vertices will correspond to $[14, 15]$. All other intervals are placed to the right of the point 15 so that they do not overlap with any of the four cliques just described.

The second batch consists of q vertices corresponding to the interval $[2, 8]$ and q vertices corresponding to the interval $[7, 13]$. All of these $2q$ intervals overlap with each other and with intervals of all colors in $\mathcal{C}_1 \cup \mathcal{C}_2'$. Thus, the algorithm uses at least $4q$ colors.

No clique is larger than $2q$ vertices, so OPT uses $2q$ colors. Since q can be arbitrarily large, no deterministic online algorithm can be better than 2-competitive, even when considering the asymptotic competitive ratio.   □

Known Interval Representation. We now assume that the vertices are revealed to the algorithm together with their interval representation. For this case, we show an improved competitive ratio of $\frac{3}{2}$. The proof of the following lower bound is a special case of the lower bound proof of Kierstead and Trotter [15].

Lemma 3. *For the problem of 2-batch coloring of interval graphs with known interval representation, no algorithm can achieve a competitive ratio strictly smaller than $\frac{3}{2}$.*

For the matching upper bound, we give a strictly $\frac{3}{2}$-competitive algorithm, called TWOBATCHES, using Algorithm FB to color the first batch of intervals and Algorithm SB to color the second batch. Intervals can be open, closed, or semi-closed. Let ω denote the maximum clique size in the full graph consisting of intervals from both batches. For any interval I, let color(I) denote the color assigned to I by TWOBATCHES. Similarly, for a set $\mathcal{I}$ of intervals, color($\mathcal{I}$) denotes the set of colors used to color the intervals in $\mathcal{I}$.

Each endpoint of a first batch interval I is called an *event point*, and this event point is associated with I. If there is a point that is an endpoint of several intervals, we have multiple copies of this point as event points each of which is associated with a different interval. We define a total order, T, on the event points. If $p < p'$, then p appears before p' in T. For the case $p = p'$, there are several cases; see the full paper [5] for details.

First Batch. It is well-known that one can color an interval graph with a maximum clique size of ω using ω colors, by maintaining a set of available colors, and traversing the event points according to the total order T: Each time a left endpoint is considered, we color its interval with a color in the set of available colors (removing it from this set); each time a right endpoint is considered, its interval's color is returned to the set of available colors. One often considers the First-Fit rule of using the minimum color in the set of available colors as a tie-breaking rule when the set of available colors contains more than one color. However, in order to establish the improved bound of $\frac{3}{2}$ on the strict competitive ratio (or even for the competitive ratio) of the algorithm for two batches, we need to use a different tie-breaking rule, the one defined by using a stack.

Algorithm FB processes the event points in the order given by T, using a stack ordering for the colors. When a right endpoint is processed, we say that the color of the associated interval is *released* and *available* until it is used again. When processing a left endpoint, the associated interval is colored with the most recently released available color (or a new color, if necessary).

For ease of presentation, we insert 2ω dummy intervals into the first batch: one clique of size ω *before* all input intervals and one clique of size ω *after* all input intervals. Since these dummy cliques do not overlap with any other intervals, each will be colored with the colors $1, 2, \ldots, \omega$, and they will not influence the behavior of Algorithm FB on the rest of the first-batch intervals.

In the following, *Maximal cliques* always refer only to first-batch intervals. For each maximal clique, we choose a point, called a *clique point*, contained in all intervals of the clique. If a clique point p appears to the right of another clique point q, we say that the clique corresponding to p appears to the right of the clique corresponding to q, and vice versa.

For each maximal clique, $\mathcal{I}$, we order the intervals of the clique by left and right endpoints, respectively, resulting in two orderings, $L_{\mathcal{I}}(\cdot)$ and $R_{\mathcal{I}}(\cdot)$. The further an endpoint is from the clique point of $\mathcal{I}$, the earlier the interval appears in the ordering. More precisely, for each interval $I \in \mathcal{I}$, $L_{\mathcal{I}}(I) = i$, if the left endpoint of I appears as the ith in T among the endpoints associated with intervals in $\mathcal{I}$. Similarly, $R_{\mathcal{I}}(I) = j$, if the right endpoint of I appears as the jth last in T among the endpoints associated with intervals in $\mathcal{I}$. As an example, consider the clique $\mathcal{I}$ consisting of the three intervals $a = [1, 6]$, $b = [2, 4]$, and $c = [3, 5]$. For this clique, we have $L_{\mathcal{I}}(a) = 1$, $L_{\mathcal{I}}(b) = 2$, $L_{\mathcal{I}}(c) = 3$ and $R_{\mathcal{I}}(a) = 1$, $R_{\mathcal{I}}(b) = 3$, $R_{\mathcal{I}}(c) = 2$.

Lemma 4. *Consider a maximal clique, $\mathcal{I}_\ell$, of size m and an interval $I_\ell \in \mathcal{I}_\ell$ such that $R_{\mathcal{I}_\ell}(I_\ell) = h$. Let $\mathcal{I}_\ell^h = \{I \in \mathcal{I}_\ell \mid R_{\mathcal{I}_\ell}(I) < h\}$ be the $h - 1$ intervals in $\mathcal{I}_\ell$ with the rightmost right endpoints. Let $\mathcal{I}_r$ be the first maximal clique of size at least h to the right of $\mathcal{I}_\ell$ and let $I_r \in \mathcal{I}_r$ be such that $L_{\mathcal{I}_r}(I_r) = h$. Finally, let p_ℓ be the right endpoint of I_ℓ, let p_r be the left endpoint of I_r, and consider the set $\mathcal{I}'$ of first-batch intervals containing a point p with $p_\ell < p < p_r$ or an endpoint p with $p_\ell <_T p <_T p_r$. Then, $color(\mathcal{I}') \subseteq color(\mathcal{I}_\ell^h)$.*

Second Batch. We now describe the algorithm, Algorithm SB, given in pseudo-code below, for coloring the second batch intervals.

A *chain* is a set of nonoverlapping second batch intervals. The algorithm starts with partitioning the second-batch intervals into ω chains (some of which may be empty). This is clearly possible, since the graph is ω-colorable.

The second batch intervals are colored in iterations, two chains per iteration. The algorithm keeps a counter, i, which is incremented once in each iteration, and maintains the set BATCH$_2$-COLORED of second batch intervals that the algorithm has already colored. In each iteration, a set of nonoverlapping first-batch intervals is *processed*. The algorithm maintains the invariant that, at the beginning of each iteration, any maximal first-batch clique of size h contains exactly $\min\{h, \omega - i\}$ unprocessed intervals.

A first-batch maximal clique of size at least $\omega - i + 1$ as well as its clique point is said to be *active*. The part of the real line between two neighboring active clique points is called a *region*. Throughout the execution of Algorithm SB, the number of regions is nondecreasing, and whenever a region is split, the chains of the region are also split by a simple projection onto each region and each resulting region will contain its boundary active clique points (in particular, this means that active clique points may belong to two regions). In each iteration, each region and its chains are treated separately.

The algorithm maintains the invariant that no uncolored second batch interval overlaps with more than one region. This is the key property, allowing the algorithm to consider one region at a time in a given iteration of the algorithm. First-batch intervals overlapping with more than one region will be cut into more intervals, with a cutting point at each active clique point contained in the interval. Thus, by cutting the intervals of an active clique of size h, the clique is replaced by two cliques of size h in neighboring regions. When a first-batch interval is cut into parts, the different parts of the interval may be processed in different iterations, but no new event points are introduced.

In the ith iteration, one chain in each region is colored with the color of a first-batch interval in the region being processed in this iteration, and another chain of the region will be colored with the color $\omega + i$, which has not been used in the region before. For any point p, let d_p be the number of second batch intervals containing p. We say that p is *covered* by a set S of intervals, if there are $\min\{d_p, i\}$ second batch intervals in S containing p.

Next, we define a set $\mathcal{P}$ of *representative points*, such that each interval between two neighboring clique points is represented by one point; see the full paper [5] for details. For a region R, we denote by $\mathcal{P}_R$ the set of representative points contained in region R (that is, $\mathcal{P}_R = R \cap \mathcal{P}$).

We use the following loop invariant for each region to establish that the algorithm TwoBatches is correct and strictly 3/2-competitive. The proof of the invariant I is based on induction on the value of i.

Invariant I:

(I1) All points p are covered by the set Batch$_2$-Colored.
(I2) No color used for an unprocessed first-batch interval contained in a region R has been used for a second batch interval intersecting region R so far.
(I3) Each active clique has exactly $\omega - i$ unprocessed intervals.
(I4) For each region R, Chain$_R$ has at most $\omega - 2i$ chains.

We use the invariant I to prove that for any input σ, TwoBatches produces a proper coloring using at most $\left\lfloor \frac{3}{2}\mathrm{OPT}(\sigma) \right\rfloor$ colors (see the full paper [5]). Combining this result with Lemma 3 shows that the optimal (strict) competitive ratio for the problem is $\frac{3}{2}$:

Theorem 3. TwoBatches *has a strict competitive ratio of* $\frac{3}{2}$.

4   Sum Coloring of Graphs in Multiple Batches

We study two cases separately: the case where the number of batches is known to the algorithm from the beginning, and the case where it is not. Once again, our lower bounds are for the competitive ratio and our upper bounds are for the strict competitive ratio.

Algorithm. SB: Coloring the second batch intervals.

1: Mark all first-batch intervals as unprocessed
2: Create an optimal coloring of the second-batch intervals, using a set $\mathcal{C}$ of ω colors
3: $R \leftarrow (-\infty, \infty)$ // *Initially, there is only one region*
4: CHAINS$_R \leftarrow \emptyset$
5: $\mathcal{P}_R \leftarrow$ the set of representative points in region R
6: **for** each color $c \in \mathcal{C}$ **do**
7:     CHAINS$_R \leftarrow$ CHAINS$_R \cup \{\{I \mid I$ is a second batch interval with color $c\}\}$
8: BATCH$_2$-COLORED $\leftarrow \emptyset$ // *Set of colored second batch intervals*
9: $i \leftarrow 0$
10: **while** $i < \lfloor \omega/2 \rfloor$ **do** // *Invariant I*
11:     // Color two chains:
12:     $i \leftarrow i + 1$
13:     *Split all regions (incl. the assoc. chains and sets of repr. points) at all active clique points*
14:     **for** *each region R containing at least one nonempty chain* **do**
15:         (CHAIN$_1$, CHAIN$_2$) $\leftarrow$ CREATECHAINS*(R)* // *See Algorithm* CREATECHAINS
16:         *// Color intervals in* CHAIN$_1$ *and* CHAIN$_2$ *using a first batch color and a new color:*
17:         $I_\ell \leftarrow$ *the unprocessed first-batch interval of the earliest event point in R*
18:         $I_r \leftarrow$ *the unprocessed first-batch interval of the latest event point in R*
19:         *Mark I_ℓ and I_r as processed*
20:         *Give all intervals in* CHAIN$_1$ *the color of I_ℓ*
21:         *Give all intervals in* CHAIN$_2$ *the color $\omega + i$*
22:         BATCH$_2$-COLORED $\leftarrow$ BATCH$_2$-COLORED $\cup$ CHAIN$_1 \cup$ CHAIN$_2$
23:         CHAIN$_R \leftarrow$ CHAIN$_R \setminus \{$CHAIN$_1$, CHAIN$_2\}$
24: // *If ω is odd, each region may have one chain left to color:*
25: **for** *each region R where* CHAINS$_R$ *contains a nonempty chain* CHAIN **do**
26:     $I \leftarrow$ *the unprocessed first-batch interval with the earliest event point in R*
27:     *Give the intervals of* CHAIN *the color of I*

Number of Batches Known in Advance. We start our study of sum coloring by examining the case where the algorithm knows the number of batches k in advance. Recall that we do not require that algorithms used within one batch be polynomial time.

Lemma 5. *There is a strictly k-competitive algorithm for sum coloring in k batches, if k is known in advance.*

Proof. For each batch, the algorithm, k-BATCHCOLOR, applies an optimal procedure, COLOR, to compute an optimal sum coloring for the subgraph induced by the set of vertices of batch i, separately from previous batches. In order to construct the solution of the input graph, k-BATCHCOLOR applies the following transformation: For every vertex v of batch i, if COLOR colors v with color c, then k-BATCHCOLOR colors v using color $f(i, c) = k \cdot (c - 1) + i$. This function f satisfies $f(i, c) \equiv i \pmod{k}$, so if $f(i, c) = f(i', c')$, for some $1 \leq i, i' \leq k$, then $i = i'$. Moreover, if $f(i, c) = f(i, c')$, then $k(c - c') = 0$, and therefore $c = c'$. Thus, vertices of different batches have different colors, and two vertices of the

---

Algorithm. CREATECHAINS(R)

1: CHAIN$_1$ ← a chain in CHAINS$_R$ containing the leftmost left endpoint
2: CHAIN$_2$ ← any other chain from CHAINS$_R$
3: **while** some point in $\mathcal{P}_R$ is *not* covered by BATCH$_2$-COLORED $\cup$ CHAIN$_1$ $\cup$ CHAIN$_2$
   do
4: p ← the leftmost point in $\mathcal{P}_R$ *not* covered by BATCH$_2$-COLORED$\cup$CHAIN$_1$$\cup$CHAIN$_2$
5: CHAIN$_3$ ← a chain from CHAINS$_R$ containing p
6: **if** for all points $q < p$ in $\mathcal{P}_R$, q is contained in CHAIN$_3$ **or** in both CHAIN$_1$ and
    CHAIN$_2$ **then**
7:    CHAIN$_2$ ← CHAIN$_3$ // CHAIN$_2$ *now refers to the chain in* CHAINS$_R$ *that* CHAIN$_3$
      refers to
8: **else**
9:    q ← the rightmost point in $\mathcal{P}_R$ left of p violating the condition
10:   CHAIN ← one of CHAIN$_1$ or CHAIN$_2$ not containing q // CHAIN *now refers to*
      a chain in CHAINS$_R$
11:   // *Do a crossover of* CHAIN *and* CHAIN$_3$ *at the point* q, *modifying* CHAIN *and*
      CHAIN$_3$ *in* CHAINS$_R$:
12:   TAIL ← $\{I \in$ CHAIN $\mid I$ starts to the right of $q\}$
13:   TAIL$_3$ ← $\{I \in$ CHAIN$_3$ $\mid I$ starts to the right of $q\}$
14:   CHAIN ← (CHAIN $\setminus$ TAIL) $\cup$ TAIL$_3$
15:   CHAIN$_3$ ← (CHAIN$_3$ $\setminus$ TAIL$_3$) $\cup$ TAIL
16: **return** (CHAIN$_1$, CHAIN$_2$)

---

same batch have the same color after the transformation if and only if they had the same color in the solution returned by COLOR. As any proper coloring of the graph provides proper colorings for the k induced subgraphs, the total cost of the k outputs of COLOR does not exceed the cost of an optimal coloring of the entire graph. For any color c and batch i, $f(i, c) \leq k \cdot c$. Thus, the cost of the output is at most k times the total cost of the k solutions returned by COLOR (for the k vertex disjoint induced subgraphs). □

We prove a matching lower bound for this case, which holds even for the asymptotic competitive ratio (see the full paper [5]). Combining that result and Lemma 5 gives the following result:

Theorem 4. *For sum coloring in k batches, with k known in advance, the optimal (strict) competitive ratio is k.*

Theorem 5. *For sum coloring of trees in k batches, First-Fit is strictly 2-competitive, and this is the best possible competitive ratio, even if k is known in advance.*

Number of Batches Unknown in Advance. Next, we consider the case where the number of batches k is not known in advance. Thus, to obtain a given competitive ratio, this ratio must be obtained after each batch. Note that the algorithm described in the proof of Lemma 5 cannot be used in this case. While the algorithm is not well defined if k is unknown in advance to the algorithm, it may seem that modifying the value of k by doubling would result in a competitive ratio of $O(k)$, but no such algorithm exists. We prove that for any positive nondecreasing sequence $f(i)$, which is defined for integer values of i (where $f(i) \geq 1$ for $i \geq 1$), no algorithm with competitive ratio $O(f(k))$ can be given if the series $S_f = \sum_{i=1}^{\infty} \frac{1}{f(i)}$ is divergent. On the other hand, we show that if this series is convergent, then such an algorithm can be given. This shows, in particular, that the best possible competitive ratio is $O(k \log k (\log \log k)^2)$ (since the series for this function converges according to the Cauchy condensation test), and it is $\Omega(k \log k \log \log k)$ (since the series for this function diverges according to the Cauchy condensation test). In fact it is $O(k \log k \log \log k \cdots (\log^{(x)} k)^2)$ and $\Omega(k \log k \log \log k \cdots \log^{(x)} k)$, for any positive integer x.

Consider a sequence $f(i)$ for which S_f is convergent, and let c_f be its limit. We present an algorithm, BATCHCOLOR$_f$, for this variant of sum coloring. Initially, all colors are declared *available*. When coloring the ith batch, its induced subgraph is first colored using an optimal procedure, COLOR. Let t_i denote the maximum color used by COLOR for batch i. For each $j = 1, 2, \ldots, t_i$ in increasing order, vertices that COLOR gives color j will be colored using the largest available color among the colors $1, 2, \ldots, \lfloor j \cdot c_f \cdot f(i) \rfloor$. Then, this color is declared *taken*. This color is now unavailable for vertices of future batches and for vertices of the current batch that were assigned a color larger than j by COLOR. If this process is successful (there always exists an available color), then we say that batch i is *feasible*.

Assuming that all batches are feasible, using arguments similar to those used for Lemma 5, we obtain an upper bound on the competitive ratio of BATCHCOLOR$_f$ as follows. Since a color used by COLOR in a particular batch is assigned to an available color by BATCHCOLOR$_f$, if all batches are feasible, each pair, (i, j), where i is a batch number and j is a color assigned by COLOR in batch i, is given a different color. Since COLOR produces a proper coloring, BATCHCOLOR$_f$ does too. The function f is nondecreasing, so the color assigned to a given vertex by BATCHCOLOR$_f$ is at most $c_f \cdot f(k)$ times the color assigned by COLOR.

Lemma 6. *Consider sum coloring in k batches, where the value of k is not known in advance. If for all $1 \leq i \leq k$, batch i is feasible, then the competitive ratio of BATCHCOLOR$_f$ is at most $c_f \cdot f(k)$.*

Lemma 7. *All batches for the algorithm BATCHCOLOR$_f$ are feasible.*

By Lemmas 6 and 7, we obtain:

Theorem 6. *Consider sum coloring in at most k batches and let f be any nondecreasing function with $f(i) \geq 1$ for all $i \geq 1$, whose series S_f converges to c_f. Then, the algorithm BATCHCOLOR$_f$ is $(c_f \cdot f(k))$-competitive, even if the value k is not known in advance.*

Now, we provide the lower bound.

Theorem 7. *Consider sum coloring in k batches, where the value of k is not known in advance. Let $f(i)$ be a nondecreasing sequence with $f(i) \geq 1$ for all $i \geq 1$, whose series S_f is divergent. Then, there is no constant c such that a competitive ratio of at most $c \cdot f(k)$ can be obtained for all $k \geq 1$.*

Proof. Assume for the sake of contradiction that there exists a constant $c > 1$ and an algorithm A, such that A is $(c \cdot f(k))$-competitive, for any number $k \geq 1$ of batches. Let $C = \max\{2c, 10\}$. Let k be such that $\sum_{i=1}^{k} 1/f(i) > 11C$ (where k must exist as the series S_f is divergent). Fix a large integer M, such that $M > 130 \cdot C^2 \cdot f(k)^2$. We say that a color a is *small* if $a \leq 10CM$.

We now describe an adversarial input. Batch i of the input consists of M^{i-1} cliques of size $3\lfloor M/f(i) \rfloor$. There are no edges between vertices in different cliques of the same batch. A vertex that A colors with a small color is called a *cheap* vertex. For each batch i, if there is at least one clique containing at least $M/f(i)$ cheap vertices, then one such clique is chosen, and the cheap vertices of this clique are called *special* vertices. In each batch, all vertices are connected to all special vertices of previous batches and to no other vertices in previous batches. Thus, no colors used for special vertices can be used in later batches, and there is at most one special vertex for each small color.

The input will contain at most k batches. If, after some batch $i < k$, the sum of colors used by A is larger than $c \cdot f(i)$ times the optimal sum of colors, there will be no more batches. Otherwise, all k batches are given. Thus, if there are fewer than k batches, the theorem trivially follows. Below, we consider the case where there are exactly k batches.

We first give an upper bound on the optimal sum of colors for the first i batches, for $1 \leq i \leq k$.

Claim 1. For every value of i (such that $1 \leq i \leq k$), the optimal sum of colors for the first i batches is at most $19M^{i+1}/(f(i))^2$.

We now show that, by the assumption that A is $(c \cdot f(i))$-competitive on i batches, $1 \leq i \leq k$, each batch i must have a clique with at least $M/f(i)$ cheap vertices. Assume for the sake of contradiction that some batch i does not contain a clique with at least $M/f(i)$ cheap vertices. Then, each clique in the batch contains at most $\lfloor M/f(i) \rfloor$ cheap vertices and hence at least $2\lfloor M/f(i) \rfloor$ vertices with colors larger than $10CM$. Thus, the sum of colors used for this batch is more than $M^{i-1} \cdot 2\lfloor M/f(i) \rfloor \cdot 10CM > 10CM^{i+1}/f(i) \geq 20cM^{i+1}/f(i)$. By Claim 1, this gives a ratio of more than

$$\frac{20cM^{i+1}/f(i)}{19M^{i+1}/(f(i))^2} > c \cdot f(i).$$

Thus, the total number of special vertices is at least $\sum_{i=1}^{k} M/f(i) > 11CM$, contradicting the fact that there is at most one special vertex for each of the small colors. $\square$

References

1. Balogh, J., Békési, J., Dósa, G., Galambos, G., Tan, Z.: Lower bound for 3-batched bin packing. Discrete Optim. **21**, 14–24 (2016)
2. Balogh, J., Békési, J., Galambos, G., Markót, M.C.: Improved lower bounds for semi-online bin packing problems. Computing **84**(1–2), 139–148 (2009)
3. Bar-Noy, A., Bellare, M., Halldorsson, M., Shachnai, H., Tamir, T.: On chromatic sums and distributed resource allocation. Inf. Comput. **140**, 183–202 (1998)
4. Borodin, A., Ivan, I., Ye, Y., Zimny, B.: On sum coloring and sum multi-coloring for restricted families of graphs. Theoret. Comput. Sci. **418**, 1–13 (2012)
5. Boyar, J., Epstein, L., Favrholdt, L.M., Larsen, K.S., Levin, A.: Batch coloring of graphs. Technical report arXiv:1610.02997 [cs.DS], arXiv (2016)
6. Dósa, G.: Batched bin packing revisited. J. Sched. (2015, in press)
7. Epstein, L.: More on batched bin packing. Oper. Res. Lett. **44**, 273–277 (2016)
8. Grötschel, M., Lovász, L., Schrijver, A.: The ellipsoid method and its consequences in combinatorial optimization. Combinatorica **1**(2), 169–197 (1981)
9. Gutin, G., Jensen, T., Yeo, A.: Batched bin packing. Discrete Optim. **2**, 71–82 (2005)
10. Gyárfás, A., Lehel, J.: On-line and first-fit colorings of graphs. J. Graph Theor. **12**, 217–227 (1988)
11. Gyárfás, A., Lehel, J.: First fit and on-line chromatic number of families of graphs. Ars Combinatoria **29**(C), 168–176 (1990)
12. Halldórsson, M.M.: Online coloring known graphs. Electron. J. Comb. **7** (2000)
13. Halldórsson, M.M., Szegedy, M.: Lower bounds for on-line graph coloring. Theoret. Comput. Sci. **130**(1), 163–174 (1994)
14. Kierstead, H.A., Smith, D.A., Trotter, W.T.: First-fit coloring on interval graphs has performance ratio at least 5. Eur. J. Comb. **51**, 236–254 (2016)
15. Kierstead, H.A., Trotter, W.T.: An extremal problem in recursive combinatorics. Congr. Numer. **33**, 143–153 (1981)
16. Kubicka, E.: The chromatic sum of a graph: history and recent developments. Int. J. Math. Math. Sci. **2004**(30), 1563–1573 (2004)
17. Kubicka, E., Schwenk, A.J.: An introduction to chromatic sums. In: 17th ACM Computer Science Conference, pp. 39–45. ACM Press (1989)
18. Narayanaswamy, N.S., Babu, R.S.: A note on first-fit coloring of interval graphs. Order **25**(1), 49–53 (2008)
19. Nicolosoi, S., Sarrafzadeh, M., Song, X.: On the sum coloring problem on interval graphs. Algorithmica **23**(2), 109–126 (1999)
20. Zhang, G., Cai, X., Wong, C.K.: Scheduling two groups of jobs with incomplete information. J. Syst. Sci. Syst. Eng. **12**, 73–81 (2003)

New Integrality Gap Results for the Firefighters Problem on Trees

Parinya Chalermsook[1,2] and Daniel Vaz[2,3(⊠)]

[1] Aalto University, Espoo, Finland
[2] Max-Planck-Institut für Informatik, Saarbrücken, Germany
{parinya,ramosvaz}@mpi-inf.mpg.de
[3] Graduate School of Computer Science, Saarland University,
Saarbrücken, Germany

Abstract. In the firefighter problem on trees, we are given a tree $G = (V, E)$ together with a vertex $s \in V$ where the fire starts spreading. At each time step, the firefighters can pick one vertex while the fire spreads from burning vertices to all their neighbors that have not been picked. The process stops when the fire can no longer spread. The objective is to find a strategy that maximizes the total number of vertices that do not burn. This is a simple mathematical model, introduced in 1995, that abstracts the spreading nature of, for instance, fire, viruses, and ideas. The firefighter problem is NP-hard and admits a $(1 - 1/e)$ approximation via LP rounding. Recently, a PTAS was announced in [1].(The $(1 - 1/e)$ approximation remained the best until very recently when Adjiashvili et al. [1] showed a PTAS. Their PTAS does not bound the LP gap.)

The goal of this paper is to develop better understanding on the power of LP relaxations for the firefighter problem. We first show a matching lower bound of $(1 - 1/e + \epsilon)$ on the integrality gap of the canonical LP. This result relies on a powerful *combinatorial gadget* that can be used to derive integrality gap results in other related settings. Next, we consider the canonical LP augmented with simple additional constraints (as suggested by Hartke). We provide several evidences that these constraints improve the integrality gap of the canonical LP: (i) Extreme points of the new LP are integral for some known tractable instances and (ii) A natural family of instances that are bad for the canonical LP admits an improved approximation algorithm via the new LP. We conclude by presenting a 5/6 integrality gap instance for the new LP.

1 Introduction

Consider the following graph-theoretic model that abstracts the fire spreading process: We are given graph $G = (V, E)$ together with the source vertex s where the fire starts. At each time step, we are allowed to pick some vertices in the graph to be saved, and the fire spreads from burning vertices to their neighbors that have not been saved so far. The process terminates when the fire cannot spread any further. This model was introduced in 1995 [13] and has been

© Springer International Publishing AG 2017
K. Jansen and M. Mastrolilli (Eds.): WAOA 2016, LNCS 10138, pp. 65–77, 2017.
DOI: 10.1007/978-3-319-51741-4_6

used extensively by researchers in several fields as an abstraction of epidemic propagation.

There are two important variants of the firefighters problem. (i) In the max-imization variant (MAX-FF), we are given graph G and source s, and we are allowed to pick one vertex per time step. The objective is to maximize the num-ber of vertices that do not burn. And (ii) In the minimization variant (MIN-FF), we are given a graph G, a source s, and a terminal set $\mathcal{X} \subseteq V(G)$, and we are allowed to pick b vertices per time step. The goal is to save all terminals in $\mathcal{X}$, while minimizing the budget b.

In this paper, we focus on the MAX-FF problem. The problem is $n^{1-\epsilon}$ hard to approximate in general graphs [2], so there is no hope to obtain any reasonable approximation guarantee. Past research, however, has focused on sparse graphs such as trees or grids. Much better approximation algorithms are known on trees: The problem is NP-hard [15] even on trees of degree at most three, but it admits a $(1 - 1/e)$ approximation algorithm. For more than a decade [2,5,6,10,14,15], there was no progress on this approximability status of this problem, until a PTAS was recently discovered [1].

Besides the motivation of studying epidemic propagation, the firefighter prob-lem and its variants are interesting due to their connections to other classical optimization problems:

- (Set cover) The firefighter problem is a special case of the *maximum coverage problem with group budget constraint (MCG)* [7]: Given a collection of sets $\mathcal{S} = \{S_1, \ldots, S_m\} : S_i \subseteq X$, together with *group constraints*, i.e. a partition of $\mathcal{S}$ into groups $G_1, \ldots, G_\ell$, we are interested in choosing one set from each group in a way that maximizes the total number of elements covered, i.e. a feasible solution is a subset $\mathcal{S}' \subseteq \mathcal{S}$ where $|\mathcal{S}' \cap G_j| \leq 1$ for every j, and $|\bigcup_{S_i \in \mathcal{S}'} S_i|$ is maximized. It is not hard to see that MAX-FF is a special case of MCG. We refer the readers to the discussion by Chekuri and Kumar [7] for more applications of MCG.
- (Cut) In a standard minimum node-cut problem, we are given a graph G together with a source-sink pair $s, t \in V(G)$. Our goal is to find a collection of nodes $V' \subseteq V(G)$ such that $G \setminus V'$ has s and t in distinct connected components. Anshelevich et al. [2] discussed that the firefighters' solution can be seen as a "cut-over-time" in which the cut must be produced gradually over many timesteps. That is, in each time step t, the algorithm is allowed to choose vertex set V'_t to remove from the graph G, and again the final goal is to "disconnect" s from t.[1] This cut-over-time problem is exactly equivalent to the minimization variant of the firefighter problem. We refer to [2] for more details about this equivalence.

1.1   Our Contributions

In this paper, we are interested in developing a better understanding of the MAX-FF problem from the perspective of LP relaxation. The canonical LP

---

[1] The notion of disconnecting the vertices here is slightly non-standard.

relaxation has been used to obtain the known $(1-1/e)$ approximation algorithm via straightforward independent LP rounding (each node is picked independently with probability proportional to its LP-value). So far, it was not clear whether an improvement was possible via this LP, for instance, via sophisticated dependent rounding schemes.[2] Indeed, for the corresponding minimization variant, MIN-FF, Chalermsook and Chuzhoy designed a dependent rounding scheme for the canonical LP in order to obtain $O(\log^* n)$ approximation algorithm, improving upon an $O(\log n)$ approximation obtained via independent LP rounding. In this paper, we are interested in studying this potential improvement for MAX-FF.

Our first result refutes such possibility for MAX-FF: we show that the integrality gap of the standard LP relaxation can be arbitrarily close to $(1 - 1/e)$.

Theorem 1. *For any $\epsilon > 0$, there is an instance (G, s) (whose size depends on ϵ) such that the ratio between optimal integral solution and fractional one is at most $(1 - 1/e + \epsilon)$.*

Our techniques rely on a powerful *combinatorial gadget* that can be used to prove integrality gap results in some other settings studied in the literature. In particular, in the b-MAX-FF problem, the firefighters can pick up to b vertices per time step, and the goal is to maximize the number of saved vertices. We provide an integrality gap of $(1 - 1/e)$ for the b-MAX-FF problem for every constant $b \in \mathbb{N}$, thus matching the algorithmic result of [9]. In the setting where an input tree has degree at most $d \in [4, \infty)$, we show an integrality gap result of $(1 - 1/e + O(1/\sqrt{d}))$. The best known algorithmic result in this setting was previously a $(1 - 1/e + \Omega(1/d))$ approximation due to [14].

Motivated by the aforementioned negative results, we search for a stronger LP relaxation for the problem. We consider adding a set of valid linear inequalities, as suggested by Hartke [12]. We show the following evidences that the new LP is a stronger relaxation than the canonical LP.

- Any extreme point of the new LP is integral for the tractable instances studied by Finbow and MacGillivray [11]. In contrast, we argue that the canonical LP does not satisfy this integrality property of extreme points.
- A family of instances, capturing the integrality gap instances of Theorem 1, admits a better than $(1 - 1/e)$ approximation algorithm via the new LP.
- When the LP solution is near-integral, e.g. for half-integral solutions, the new LP is provably better than the old one.

Our results are the first rigorous evidences that Hartke's constraints lead to improvements upon the canonical LP. All the aforementioned algorithmic results exploit the new LP constraints in dependent LP rounding procedures. In particular, we propose a two-phase dependent rounding algorithm, which can be used in deriving the second and third results. We believe the new LP has an integrality gap strictly better than $(1 - 1/e)$, but we are unable to analyze it.

---

[2] Cai et al. [5] claimed an LP-respecting integrality gap of $(1-1/e)$, but many natural rounding algorithms in the context of this problem are not LP respecting, e.g. in [6].

Finally, we show a limitation of the new LP by presenting a family of instances, whose integrality gap can be arbitrarily close to $5/6$. This improves the known integrality gap ratio [12], and puts the integrality gap answer somewhere between $(1 - 1/e)$ and $5/6$. Closing this gap is, in our opinion, an interesting open question.

Organization: In Sect. 2, we formally define the problem and present the LP relaxation. In Sect. 3, we present the bad integrality gap instances. We present the LP augmented with Hartke's constraints in Sect. 4 and discuss the relevant evidences of its power in comparison to the canonical LP. Some proofs are omitted for space constraint, and are presented in the full version.

Related Results: King and MacGillivray showed that the firefighter problem on trees is solvable in polynomial time if the input tree has degree at most three, with the fire starting at a degree-2 vertex. From exponential time algorithm's perspective, Cai et al. showed $2^{O(\sqrt{n}\log n)}$ time, exact algorithm. The discrete mathematics community pays particularly high attention to the firefighter problem on grids [10,16], and there has also been some work on infinite graphs [13].

The problem also received a lot of attention from the parameterized complexity perspectives [3,5,8] and on many special cases, e.g., when the tree has bounded pathwidth [8] and on bounded degree graphs [4,8].

Recent Update: Very recently, Adjiashvili et al. [1] showed a polynomial time approximation scheme (PTAS) for the MAX-FF problem, therefore settling the approximability status. Their results, however, do not bound the LP integrality gap. We believe that the integrality gap questions are interesting despite the known approximation guarantees.

2   Preliminaries

A formal definition of the problem is as follows. We are given a graph G and a source vertex s where the fire starts spreading. A *strategy* is described by a collection of vertices $\mathcal{U} = \{u_t\}_{t=1}^n$ where $u_t \in V(G)$ is the vertex picked by firefighters at time t. We say that a vertex $u \in V(G)$ is *saved* by the strategy $\mathcal{U}$ if for each path $P = (s = v_0, \ldots, v_z = u)$ from s to u, we have $v_i \in \{u_1, \ldots, u_i\}$ for some $i = 1, \ldots, z$. A vertex v not saved by $\mathcal{U}$ is said to be a *burning vertex*. The objective of the problem is to compute $\mathcal{U}$ so as to maximize the total number of saved vertices. Denote by $\mathsf{OPT}(G, s)$ the number of vertices saved by an optimal solution.

When G is a tree, we think of G as being partitioned into layers $L_1, \ldots, L_\lambda$ where λ is the height of the tree, and L_i contains vertices whose distance is exactly i from s. Every strategy has the following structure.

Proposition 1. *Consider the firefighters problem's instance (G, s) where G is a tree. Let $\mathcal{U} = \{u_1, \ldots, u_n\}$ be any strategy. Then there is another strategy*

$\mathcal{U}' = \{u'_t\}$ where u'_t belongs to layer t in G, and $\mathcal{U}'$ saves at least as many vertices as $\mathcal{U}$ does.

We remark that this structural result holds only when G is a tree.

LP Relaxation: This paper focuses on the linear programming aspect of the problem. For any vertex v, let P_v denote the (unique) path from s to v, and let T_v denote the subtree rooted at v. A natural LP relaxation is denoted by (LP-1): We have variable x_v indicating whether v is picked by the solution, and y_v indicating whether v is saved.

(LP-1)

$$\max \sum_{v \in V} y_v$$

$$\sum_{v \in L_j} x_v \leq 1 \text{ for each layer } j$$

$$y_v \leq \sum_{u \in P_v} x_u \text{ for each } v \in V$$

$$x_v, y_v \in [0,1] \text{ for each } v$$

(LP-2)

$$\max \sum_{v \in \mathcal{X}} y_v$$

$$\sum_{v \in L_j} x_v \leq 1 \text{ for each layer } j$$

$$y_v \leq \sum_{u \in P_v} x_u \text{ for each } v \in \mathcal{X}$$

$$x_v, y_v \in [0,1] \text{ for each } v$$

Let $\mathsf{LP}(T,s)$ denote the optimal fractional LP value for an instance (T,s). The integrality gap $\mathsf{gap}(T,s)$ of the instance (T,s) is defined as $\mathsf{gap}(T,s) = \mathsf{OPT}(T,s)/\mathsf{LP}(T,s)$. The integrality gap of the LP is defined as $\inf_T \mathsf{gap}(T,s)$.

Firefighters with Terminals: We consider a more general variant of the problem, where we are only interested in saving a subset $\mathcal{X}$ of vertices, which we call *terminals*. The goal is now to maximize the number of saved terminals. An LP formulation of this problem, given an instance $(T, v, \mathcal{X})$, is denoted by (LP-2). The following lemma argues that these two variants are "equivalent" from the perspectives of LP relaxation.

Lemma 1. *Let $(T, \mathcal{X}, s)$, with $|\mathcal{X}| > 0$, be an input for the terminal firefighters problem that gives an integrality gap of γ for (LP-2), and that the value of the fractional optimal solution is at least 1. Then, for any $\epsilon > 0$, there is an instance (T', s') that gives an integrality gap of $\gamma + \epsilon$ for (LP-1).*

We will, from now on, focus on studying the integrality gap of (LP-2).

3   Integrality Gap of (LP-2)

We first discuss the integrality gap of (LP-2) for a general tree. We use the following combinatorial gadget.

Gadget: A (M, k, δ)-*good gadget* is a collection of trees $\mathcal{T} = \{T_1, \ldots, T_M\}$, with roots $r_1, \ldots, r_M$ where r_i is a root of T_i, and a subset $\mathcal{S} \subseteq \bigcup V(T_i)$ that satisfy the following properties:

- (Uniform depth) We think of these trees as having layers $L_0, L_1, \ldots, L_h$, where L_j is the union over all trees of all vertices at layer j and $L_0 = \{r_1, \ldots, r_m\}$. All leaves are in the same layer L_h.
- (LP-friendly) For any layer $L_j, j \geq 1$, we have $|\mathcal{S} \cap L_j| \leq k$ (and $|\mathcal{S} \cap L_0| = 0$). Moreover, for any tree T_i and a leaf $v \in V(T_i)$, the unique path from r_i to v must contain exactly one vertex in $\mathcal{S}$.
- (Integrally adversarial) Let $\mathcal{B} \subseteq \{r_1, \ldots, r_M\}$ be any subset of roots. Consider a subset of vertices $\mathcal{U} = \{u_j\}_{j=1}^{h}$ such that $u_j \in L_j$. For $r_i \in \mathcal{B}$ and a leaf $v \in L_h \cap V(T_i)$, we say that v is $(\mathcal{U}, \mathcal{B})$-*risky* if the unique path from r_i to v does not contain any vertex in $\mathcal{U}$. There must be at least $(1 - 1/k - \delta)\frac{|\mathcal{B}|}{M}|L_h|$ vertices in L_h that are $(\mathcal{U}, \mathcal{B})$-risky, for all choices of $\mathcal{B}$ and $\mathcal{U}$.

We say that vertices in $\mathcal{S}$ are *special* and all other vertices are *regular*.

Lemma 2. *For any integers $k \geq 2$, $M \geq 1$, and any real number $\delta > 0$, a (M, k, δ)-good gadget exists. Moreover, the gadget contains at most $(k/\delta)^{O(M)}$ vertices.*

We first show how to use this lemma to derive our final construction. The proof of the lemma follows later.

Construction: Our construction proceeds in k phases, and we will define it inductively. The first phase of the construction is simply a $(1, k, \delta)$-good gadget. Now, assume that we have constructed the instance up to phase q. Let $l_1, \ldots, l_{M_q} \in L_{\alpha_p}$ be the leaves after the construction of phase q that all lie in layer α_q. In phase $q + 1$, we take the (M_q, k, δ)-good gadget $(\mathcal{T}_q, \{r_q\}, \mathcal{S}_q)$; recall that such a gadget consists of M_q trees. For each $i = 1, \ldots, M_q$, we unify each root r_i with the leaf l_i. This completes the description of the construction.

Denote by $\bar{\mathcal{S}}_q = \bigcup_{q' \leq q} \mathcal{S}_{q'}$ the set of all special vertices in the first q phases. After phase q, we argue that our construction satisfies the following properties:

- All leaves are in the same layer α_q.
- For every layer L_j, $|L_j \cap \bar{\mathcal{S}}_q| \leq k$. For every path P from the root to $v \in L_{\alpha_i}$, $|P \cap \bar{\mathcal{S}}_q| = q$.
- For any integral solution $\mathcal{U}$, at least $|L_{\alpha_q}|((1 - 1/k)^q - q\delta)$ vertices of L_{α_q} burn.

It is clear from the construction that the leaves after phase q are all in the same layer. As to the second property, the properties of the gadget ensure that there are at most k special vertices per layer. Moreover, consider each path P from the root to some vertex $v \in L_{\alpha_{q+1}}$. We can split this path into two parts $P = P' \cup P''$ where P' starts from the root and ends at some $v' \in L_{\alpha_q}$, and P'' starts at v' and ends at v. By the induction hypothesis, $|P' \cap \bar{\mathcal{S}}_q| = q$ and the second property of the gadget guarantees that $|P'' \cap \mathcal{S}_{q+1}| = 1$.

To prove the final property, consider a solution $\mathcal{U} = \{u_1, \ldots, u_{\alpha_{q+1}}\}$, which can be seen as $\mathcal{U}' \cup \mathcal{U}''$ where $\mathcal{U}' = \{u_1, \ldots, u_{\alpha_q}\}$ and $\mathcal{U}'' = \{u_{\alpha_q+1}, \ldots, u_{\alpha_{q+1}}\}$. By the induction hypothesis, we have that at least $((1 - 1/k)^q - q\delta)|L_{\alpha_q}|$ vertices in L_{α_q} burn; denote these burning vertices by $\mathcal{B}$. The third property of the gadget will ensure that at least $(1 - 1/k - \delta)\frac{|\mathcal{B}|}{M_q}|L_{\alpha_{q+1}}|$ vertices in $L_{\alpha_{q+1}}$ must be $(\mathcal{U}'', \mathcal{B})$-risky. For each risky vertex $v \in L_{\alpha_{q+1}}$, a unique path from the root to $v' \in \mathcal{B}$ does not contain any vertex in $\mathcal{U}'$, and also the path from v' to v does not contain a vertex in $\mathcal{U}''$ (due to the fact that it is $(\mathcal{U}'', \mathcal{B})$-risky.) This implies that such vertex v must burn. Therefore, the fraction of burning vertices in layer $L_{\alpha_{q+1}}$ is at least $(1 - 1/k - \delta)|\mathcal{B}|/M_q \geq (1 - 1/k - \delta)((1 - 1/k)^q - q\delta)$, by induction hypothesis. This number is at least $(1 - 1/k)^{q+1} - (q+1)\delta$, maintaining the invariant.

After the construction of all k phases, the leaves are designated as the terminals $\mathcal{X}$. Also, $M_{q+1} \leq (k/\delta)^{2M_q}$, which means that, after k phases, M_k is at most a tower function of $(k/\delta)^2$, that is, $(k/\delta)^{2(k/\delta)^{\cdots}}$ with $k-1$ such exponentiations. The total size of the construction is $\sum_q (k/\delta)^{2M_q} \leq (k/\delta)^{2M_k} = O(M_{k+1})$.

For an example construction $(k = 2)$, refer to the full version.

Theorem 2. *A fractional solution, that assigns $x_v = 1/k$ to each special vertex v, saves every terminal. On the other hand, any integral solution can save at most a fraction of $1 - (1 - 1/k)^k + \epsilon$.*

3.1   Proof of Lemma 2

We now show that the (M, k, δ)-good gadget exists for any value of $M \in \mathbb{N}$, $k \in \mathbb{N}, k \geq 2$ and $\delta \in \mathbb{R}_{>0}$. We first describe the construction and then show that it has the desired properties.

Construction: Throughout the construction, we use a structure which we call *spider*. A spider is a tree in which every node except the root has at most one child. If a node has no children (i. e. a leaf), we call it a *foot* of the spider. We call the paths from the root to each foot the *legs* of the spider.

Let $D = \lceil 4/\delta \rceil$. For each $i = 1, \ldots, M$, the tree T_i is constructed as follows. We have a spider rooted at r_i that contains kD^{i-1} legs. Its feet are in D^{i-1} consecutive layers, starting at layer $\alpha_i = 1 + \sum_{j<i} D^{j-1}$; each such layer has k feet. Denote by $\mathcal{S}^{(i)}$ the feet of these spiders. Next, for each vertex $v \in \mathcal{S}^{(i)}$, we have a spider rooted at v, having D^{2M-i+1} feet, all of which belong to layer $\alpha = 1 + \sum_{j \leq M} D^{j-1}$. The set $\mathcal{S}$ is defined as $\mathcal{S} = \bigcup_{i=1}^{M} \mathcal{S}^{(i)}$. This concludes the construction. We will use the following observation:

Observation 1. *For each root r_i, the number of leaves of T_i is kD^{2M}.*

Analysis: We now prove that the above gadget is (M, k, δ)-good. The construction ensures that all leaves are in the same layer L_α.

The second property also follows obviously from the construction: For $i \neq i'$, we have that $\mathcal{S}^{(i)} \cap \mathcal{S}^{(i')} = \emptyset$, and that each layer contains exactly k vertices from $\mathcal{S}^{(i)}$. Moreover, any path from r_i to the leaf of T_i must go through a vertex in $\mathcal{S}^{(i)}$.

The third and final property is established by the following two lemmas.

Lemma 3. *For any $r_i \in \mathcal{B}$ and any subset of vertices $\mathcal{U} = \{u_j\}_{j=1}^h$ such that $u_j \in L_j$, a fraction of at least $(1 - 1/k - 2/D)$ of $\mathcal{S}^{(i)}$ are $(\mathcal{U}, \mathcal{B})$-risky.*

Lemma 4. *Let $v \in \mathcal{S}^{(i)}$ that is $(\mathcal{U}, \mathcal{B})$-risky. Then at least $(1 - 2/D)$ fraction of descendants of v in L_α must be $(\mathcal{U}, \mathcal{B})$-risky.*

Combining the above two lemmas, for each $r_i \in \mathcal{B}$, the fraction of leaves of T_i that are $(\mathcal{U}, \mathcal{B})$-risky are at least $(1 - 1/k - 2/D)(1 - 2/D) \geq (1 - 1/k - 4/D)$. Therefore, the total number of such leaves, over all trees in $\mathcal{T}$, are $(1 - 1/k - \delta)$ $|\mathcal{B}||L_\alpha|/M$.

We extend the construction to other settings in the full version.

4   Hartke's Constraints

Due to the integrality gap result in the previous section, there is no hope to improve the best known algorithms via the canonical LP relaxation. Hartke [12] suggested adding the following constraints to narrow down the integrality gap of the LP.

$$\sum_{u \in P_v \cup (T_v \cap L_j)} x_u \leq 1 \text{ for each vertex } v \in V(T) \text{ and layer } L_j \text{ below the layer of } v$$

We write the new LP with these constraints below:

(LP')
$$\max \sum_{v \in V} y_v$$

$$\sum_{u \in P_v \cup (T_v \cap L_j)} x_u \leq 1 \text{ for each layer } j \text{ below vertex } v$$

$$y_v \leq \sum_{u \in P_v} x_u \text{ for each } v \in V$$

$$x_v, y_v \in [0,1] \text{ for each } v$$

Proposition 2. *Given the values $\{x_v\}_{v \in V(T)}$ that satisfy the first set of constraints, then the solution (x, y) defined by $y_v = \sum_{u \in P_v} x_v$ is feasible for (LP') and at least as good as any other feasible (x, y').*

In this section, we study the power of this LP and provide three evidences that it may be stronger than (LP-1).

4.1 New Properties of Extreme Points

In this section, we show that Finbow et al. tractable instances [11] admit a polynomial time exact algorithm via (LP') (in fact, any optimal extreme point for (LP') is integral.) In contrast, we show that (LP-1) contains an extreme point that is not integral.

We first present the following structural lemma.

Lemma 5. *Let $(\mathbf{x}, \mathbf{y})$ be an optimal extreme point for (LP') on instance T rooted at s. Suppose s has two children, denoted by a and b. Then $x_a, x_b \in \{0, 1\}$.*

Finbow et al. Instances: In this instance, the tree has degree at most 3 and the root has degree 2. Finbow et al. [11] showed that this is polynomial time solvable.

Theorem 3. *Let (T, s) be an input instance where T has degree at most 3 and s has degree two. Let (x, y) be a feasible fractional solution for (LP'). Then there is a polynomial time algorithm that saves at least $\sum_{v \in V(T)} y_v$ vertices.*

Bad Instance for (LP-1): We show in Fig. 1 a Finbow et al. instance as well as a solution for (LP-1) that is optimal and an extreme point, but not integral.

Fig. 1. Instance with a non-integral extreme point for (LP-1). *Gray vertices:* $x_v = 1/2$; otherwise: $x_v = 0$.

4.2 Rounding 1/2-Integral Solutions

We say that the LP solution (x, y) is $(1/k)$-integral if, for each v, we have $x_v = r_v/k$ for some integer $r_v \in \{0, \ldots, k\}$. By standard LP theory, one can assume that the LP solution is $(1/k)$-integral for some polynomially large integer k.

In this section, we consider the case when $k = 2$ (1/2-integral LP solutions). From Theorem 2, (LP-1) is not strong enough to obtain a $3/4 + \epsilon$ approximation algorithm, for any $\epsilon > 0$. Here, we show a $5/6$ approximation algorithm based on rounding (LP').

Theorem 4. *Given a solution (x, y) for (LP') that is 1/2-integral, there is a polynomial time algorithm that produces a solution of cost $5/6 \sum_{v \in V(T)} y_v$.*

We believe that the extreme points in some interesting special cases will be 1/2-integral.

Algorithm's Description: Initially, $\mathcal{U} = \emptyset$. Our algorithm considers the layers $L_1, \ldots, L_n$ in this order. When the algorithm looks at layer L_j, it picks a vertex u_j and adds it to $\mathcal{U}$, as follows. Consider $A_j \subseteq L_j$, where $A_j = \{v \in L_j : x_v > 0\}$. Let $A'_j \subseteq A_j$ contain vertices v such that there is no ancestor of v that belongs to $A_{j'}$ for some $j' < j$, and $A''_j = A_j \setminus A'_j$, i.e. for each $v \in A''_j$, there is another vertex $u \in A_{j'}$ for some $j' < j$ such that u is an ancestor of v. We choose the vertex u_j based on the following rules:

- If there is only one $v \in A_j$, such that v is not saved by $\mathcal{U}$ so far, choose $u_j = v$.
- Otherwise, if $|A'_j| = 2$, pick u_j at random from A'_j with uniform probability. Similarly, if $|A''_j| = 2$, pick u_j at random from A''_j.
- Otherwise, we have the case $|A'_j| = |A''_j| = 1$. In this case, we pick vertex u_j from A'_j with probability $1/3$; otherwise, we take from A''_j.

4.3   Ruling Out the Gap Instances in Sect. 3

In this section, we show that the integrality gap instances for (LP-1) presented in the previous section admit a better than $(1 - 1/e)$ approximation via (LP'). To this end, we introduce the concept of well-separable LP solutions and show an improved rounding algorithm for solutions in this class.

Let $\eta \in (0, 1)$. Given an LP solution (x, y) for (LP-1) or (LP'), we say that a vertex v is η-light if $\sum_{u \in P_v \setminus \{v\}} x_u < \eta$; if a vertex v is not η-light, we say that it is η-heavy. A fractional solution is said to be η-separable if for each layer j, either all vertices in L_j are η-light, or they are all η-heavy. For an η-separable LP solution (x, y), each layer L_j is either an η-light layer that contains only η-light vertices, or η-heavy layer that contains only η-heavy vertices.

Observation 2. *The LP solution presented in Sect. 3 is η-separable for all values of $\eta \in \{1/k, 2/k, \ldots, 1\}$.*

Theorem 5. *If the LP solution (x, y) is η-separable for some η, then there is an efficient algorithm that produces an integral solution of cost $(1 - 1/e + f(\eta)) \sum_v y_v$, where $f(\eta)$ is some function depending only on η.*

Algorithm: Let T be an input tree, and (x, y) be a solution for (LP') on T that is η-separable for some constant $\eta \in (0, 1)$. Our algorithm proceeds in two phases. In the first phase, it performs randomized rounding independently for each η-light layer. Denote by V_1 the (random) collection of vertices selected in this phase. Then, in the second phase, our algorithm performs randomized rounding conditioned on the solutions in the first phase. In particular, when we process each η-heavy layer L_j, let $\tilde{L}_j$ be the collection of vertices that have not yet been saved by V_1. We sample one vertex $v \in \tilde{L}_j$ from the distribution $\left\{\frac{x_v}{x(\tilde{L}_j)}\right\}_{v \in \tilde{L}_j}$. Let V_2 be the set of vertices chosen from the second phase. This completes the description of our algorithm.

4.4   Integrality Gap for (LP')

In this section, we present an instance where (LP') has an integrality gap of $5/6 + \epsilon$, for any $\epsilon > 0$. Interestingly, this instance admits an optimal $\frac{1}{2}$-integral LP solution.

Fig. 2. Gadget used to get 5/6 integrality gap. Special vertices are colored *gray*.

Gadget: The motivation of our construction is a simple gadget represented in Fig. 2. In this instance, vertices are either *special* (colored gray) or *regular*. This gadget has three properties of our interest:

- If we assign an LP-value of $x_v = 1/2$ to every special vertex, then this is a feasible LP solution that ensures $y_u = 1$ for every leaf u.
- For any integral solution $\mathcal{U}$ that does not pick any vertex in the first layer of this gadget, at most 2 out of 3 leaves of the gadget are saved.
- Any pair of special vertices in the same layer do not have a common ancestor inside this gadget.

Our integrality gap instance is constructed by creating partially overlapping copies of this gadget. We describe it formally below.

Construction: The first layer of this instance, L_1, contains 4 nodes: two special nodes, which we name $a(1)$ and $a(2)$, and two regular nodes, which we name $b(1)$ and $b(2)$. We recall the definition of spider from Sect. 3.1.
    Let $\alpha = 5\lceil 1/\epsilon \rceil$. The nodes $b(1)$ and $b(2)$ are the roots of two spiders. Specifically, the spider Z_1 rooted at $b(1)$ has α feet, with one foot per layer, in consecutive layers $L_2, \ldots, L_{\alpha+1}$. For each $j \in [\alpha]$, denote by $b'(1, j)$, the jth foot of spider Z_1. The spider Z_2, rooted at $b(2)$, has α^2 feet, with one foot per layer, in layers $L_{\alpha+2}, \ldots, L_{\alpha^2+\alpha+1}$. For each $j \in [\alpha^2]$, denote by $b'(2, j)$, the jth foot of spider Z_2. All the feet of spiders Z_1 and Z_2 are special vertices.
    For each $j \in [\alpha]$, the node $b'(1, j)$ is also the root of spider $Z'_{1,j}$, with α^2 feet, lying in the α^2 consecutive layers $L_{2+\alpha+j\alpha^2}, \ldots, L_{1+\alpha+(j+1)\alpha^2}$ (one foot per layer). For $j' \in [\alpha^2]$, let $b''(1, j, j')$ denote the j'-th foot of spider $Z'_{1,j}$ that lies in layer $L_{1+\alpha+j\alpha^2+j'}$. Notice that we have α^3 such feet of these spiders $\{Z'_{1,j}\}_{j=1}^{\alpha}$ lying in layers $L_{2+\alpha+\alpha^2}, \ldots, L_{1+\alpha+\alpha^2+\alpha^3}$. Similarly, for each $j \in [\alpha^2]$, the node $b'(2, j)$ is the root of spider $Z'_{2,j}$ with α^2 feet, lying in consecutive layers

$L_{2+\alpha+\alpha^3+j\alpha^2}, \ldots, L_{1+\alpha+\alpha^3+(j+1)\alpha^2}$. We denote by $b''(2, j, j')$ the j'-th foot of this spider.

The special node $a(1)$ is also the root of spider W_1 which has $\alpha + \alpha^3$ feet: The first α feet, denoted by $a'(1, j)$ for $j \in [\alpha]$, are aligned with the nodes $b'(1, j)$, i.e. for each $j \in [\alpha]$, the foot $a'(1, j)$ of spider W_1 is in the same layer as the foot $b'(1, j)$ of Z_1. For each $j \in [\alpha], j' \in [\alpha^2]$, we also have a foot $a''(1, j, j')$ which is placed in the same layer as $b''(1, j, j')$. Similarly, the special node $a(2)$ is the root of spider W_2 having $\alpha^2 + \alpha^4$ feet. For $j \in [\alpha^2]$, spider W_2 has a foot $a'(2, j)$ placed in the same layer as $b'(2, j)$. For $j \in [\alpha^2], j' \in [\alpha^2]$, W_2 also has a foot $a''(2, j, j')$ in the layer of $b''(2, j, j')$. All the feet of both W_1 and W_2 are special vertices.

Finally, for $i \in \{1, 2\}$, and $j \in [\alpha^i]$, each node $a'(i, j)$ has α^{5-i} children, which are leaves of the instance. For $j \in [\alpha], j' \in [\alpha^2]$, the nodes $b''(i, j, j')$, $a''(i, j, j')$ have α^{3-i} children each which are also leaves of the instance. The set of terminals $\mathcal{X}$ is simply the set of leaves.

Proposition 3. *We have $|\mathcal{X}| = 6\alpha^5$. Moreover, (i) the number of terminals in subtrees $T_{a(1)} \cup T_{b(1)}$ is $3\alpha^5$, and (ii) the number of terminals in subtrees $T_{a(2)} \cup T_{b(2)}$ is $3\alpha^5$.*

Fractional Solution: Our construction guarantees that any path from root to leaf contains 2 special vertices: For a leaf child of $a'(i, j)$, its path towards the root must contain $a'(i, j)$ and $a(i)$. For a leaf child of $a''(i, j, j')$, its path towards the root contains $a''(i, j, j')$ and $a(i)$. For a leaf child of $b''(i, j, j')$, the path towards the root contains $b''(i, j, j')$ and $b'(i, j)$.

Lemma 6. *For each special vertex v, for each layer L_j below v, the set $L_j \cap T_v$ contains at most one special vertex.*

Notice that, there are at most two special vertices per layer. We define the LP solution x, with $x_v = 1/2$ for every special vertex v and $x_v = 0$ for all other vertices. It is easy to verify that this is a feasible solution.

Integral Solution: We argue that any integral solution cannot save more than $(1 + 5/\alpha)5\alpha^5$ terminals. The following lemma is the key to our analysis.

Lemma 7. *Any integral solution $\mathcal{U} : \mathcal{U} \cap \{a(1), b(1)\} = \emptyset$ saves at most $(1 + 5/\alpha)5\alpha^5$ terminals.*

Lemma 8. *Any integral solution $\mathcal{U} : \mathcal{U} \cap \{a(2), b(2)\} = \emptyset$ saves at most $(1 + 5/\alpha)5\alpha^5$ terminals.*

Since nodes $a(1)$, $a(2)$, $b(1)$, $b(2)$ are in the first layer, it is only possible to save one of them. Therefore, either Lemma 7 or Lemma 8 apply, which concludes the analysis.

5   Conclusion and Open Problems

In this paper, we settled the integrality gap question for the standard LP relaxation. Our results ruled out the hope to use the canonical LP to obtain better approximation results. While a recent paper settled the approximability status of the problem [1], the question whether an improvement over $(1 - 1/e)$ can be done via LP relaxation is of independent interest. We provide some evidences that Hartke's LP is a promising candidate for doing so. Another interesting question is to find a more general graph class that admits a constant approximation algorithm. We believe that this is possible for bounded treewidth graphs.

References

1. Adjiashvili, D., Baggio, A., Zenklusen, R.: Firefighting on Trees Beyond Integrality Gaps. ArXiv e-prints, January 2016
2. Anshelevich, E., Chakrabarty, D., Hate, A., Swamy, C.: Approximability of the firefighter problem - computing cuts over time. Algorithmica **62**(1–2), 520–536 (2012)
3. Bazgan, C., Chopin, M., Cygan, M., Fellows, M.R., Fomin, F.V., van Leeuwen, E.J.: Parameterized complexity of firefighting. JCSS **80**(7), 1285–1297 (2014)
4. Bazgan, C., Chopin, M., Ries, B.: The firefighter problem with more than one firefighter on trees. Discrete Appl. Math. **161**(7–8), 899–908 (2013)
5. Cai, L., Verbin, E., Yang, L.: Firefighting on trees: $(1-1/e)$-approximation, fixed parameter tractability and a subexponential algorithm. In: ISAAC (2008)
6. Chalermsook, P., Chuzhoy, J.: Resource minimization for fire containment. In: SODA (2010)
7. Chekuri, C., Kumar, A.: Maximum coverage problem with group budget constraints and applications. In: Jansen, K., Khanna, S., Rolim, J.D.P., Ron, D. (eds.) APPROX/RANDOM -2004. LNCS, vol. 3122, pp. 72–83. Springer, Heidelberg (2004). doi:10.1007/978-3-540-27821-4_7
8. Chlebíková, J., Chopin, M.: The firefighter problem: a structural analysis. In: Cygan, M., Heggernes, P. (eds.) IPEC 2014. LNCS, vol. 8894, pp. 172–183. Springer, Heidelberg (2014). doi:10.1007/978-3-319-13524-3_15
9. Costa, V., Dantas, S., Dourado, M.C., Penso, L., Rautenbach, D.: More fires and more fighters. Discrete Appl. Math. **161**(16–17), 2410–2419 (2013)
10. Develin, M., Hartke, S.G.: Fire containment in grids of dimension three and higher. Discrete Appl. Math. **155**(17), 2257–2268 (2007)
11. Finbow, S., MacGillivray, G.: The firefighter problem: a survey of results, directions and questions. Australas. J. Comb. **43**, 57–77 (2009)
12. Hartke, S.G.: Attempting to narrow the integrality gap for the firefighter problem on trees. Discrete Methods Epidemiol. **70**, 179–185 (2006)
13. Hartnell, B.: Firefighter! an application of domination. In: Manitoba Conference on Combinatorial Mathematics and Computing (1995)
14. Iwaikawa, Y., Kamiyama, N., Matsui, T.: Improved approximation algorithms for firefighter problem on trees. IEICE Trans. **94–D**(2), 196–199 (2011)
15. King, A., MacGillivray, G.: The firefighter problem for cubic graphs. Discrete Math. **310**(3), 614–621 (2010)
16. Wang, P., Moeller, S.A.: Fire control on graphs. J. Comb. Math. Comb. Comput. **41**, 19–34 (2002)

A Multiplicative Weights Update Algorithm for Packing and Covering Semi-infinite Linear Programs

Khaled Elbassioni[1], Kazuhisa Makino[2], and Waleed Najy[1(✉)]

[1] Masdar Institute of Science and Technology, P.O. Box 54224,
Abu Dhabi, UAE
{kelbassioni,wnajy}@masdar.ac.ae
[2] Research Institute for Mathematical Sciences (RIMS), Kyoto University,
Kyoto 606-8502, Japan
makino@kurims.kyoto-u.ac.jp

Abstract. We consider the following semi-infinite linear programming problems: max (resp., min) $c^T x$ s.t. $y^T A_i x + (d^i)^T x \leq b_i$ (resp., $y^T A_i x + (d^i)^T x \geq b_i$), for all $y \in \mathcal{Y}_i$, for $i = 1, \ldots, N$, where $\mathcal{Y}_i \subseteq \mathbb{R}_+^{m_i}$ are given compact convex sets and $A_i \in \mathbb{R}_+^{m_i \times n}$, $b = (b_1, \ldots, b_N) \in \mathbb{R}_+^N$, $d_i \in \mathbb{R}_+^n$, and $c \in \mathbb{R}_+^n$ are given non-negative matrices and vectors. This general framework is useful in modeling many interesting problems. For example, it can be used to represent a sub-class of Robust optimization in which the coefficients of the constraints are drawn from convex uncertainty sets $\mathcal{Y}_i$, and the goal is to optimize the objective function for the worst-case choice in each $\mathcal{Y}_i$. When the uncertainty sets $\mathcal{Y}_i$ are ellipsoids, we obtain a sub-class of Second-Order Cone Programming. We show how to extend the multiplicative weights update method to derive approximation schemes for the above packing and covering problems. When the sets $\mathcal{Y}_i$ are simple, such as ellipsoids or boxes, this yields substantial improvements in the running time over general convex programming solvers.

Keywords: Multiplicative weights update · Robust optimization · Second-order cone programming · Packing and covering

1 Introduction

1.1 Problem Definition and Related Work

We consider the following semi-infinite linear programming problems:

$$\max_{x \in \mathbb{R}_+^n} c^T x \text{ s.t. } y^T A_i x + (d^i)^T x \leq b_i, \forall y \in \mathcal{Y}_i, \ i \in [N], \qquad \text{(PACKING)}$$

$$\min_{x \in \mathbb{R}_+^n} c^T x \text{ s.t. } y^T A_i x + (d^i)^T x \geq b_i, \forall y \in \mathcal{Y}_i, \ i \in [N], \qquad \text{(COVERING)}$$

© Springer International Publishing AG 2017
K. Jansen and M. Mastrolilli (Eds.): WAOA 2016, LNCS 10138, pp. 78–91, 2017.
DOI: 10.1007/978-3-319-51741-4_7

where, for $i = 1, \ldots, N$, $\mathcal{Y}_i \subseteq \mathbb{R}_+^{m_i}$ are compact convex sets, $b := (b_1, \ldots, b_N) \in \mathbb{R}_+^N$, $d^i, c \in \mathbb{R}_+^n$, are non-negative vectors, and $A_i \in \mathbb{R}_+^{m_i \times n}$ are non-negative matrices. As we shall see, the above formulation is general enough to capture many interesting problems. For example, it can be used to represent the class of *robust packing and covering linear programs* in which the coefficients of the constraints are drawn from *convex uncertainty sets* $\mathcal{Y}_i$. A particularly important case is when the sets $\mathcal{Y}_i$ are ellipsoids, in which case we obtain a sub-class of *Second-Order Cone Programming* (SOCP). Robust optimization and SOCP are important tools in many areas such as machine learning [28], portfolio optimization [14], supply chain management [5], and many other applications [4].

While there is extensive literature on solving robust optimization problems and SOCP, mainly based on *interior point methods* [1,16], most (if not all) of the existing algorithms do not take advantage of the special structure of the constraints in (PACKING) and (COVERING), which is implied by the non-negativity of these constraints. Recently, there has been growing interest in finding simpler and faster approximation algorithms for convex optimization problems, sacrificing the dependence on the approximation accuracy ϵ from polylog($\frac{1}{\varepsilon}$) to poly($\frac{1}{\varepsilon}$) in exchange of efficiency in terms of other input parameters; see, e.g., [2,12,13,17,18,21]. One of the simplest and most widely used methods is the *multiplicative weights update* (MWU) method, which has its roots in game theory [7,27], and was rediscovered more formally in the machine learning community [11,22], and in several other areas. It can be thought of as a learning algorithm which, at a very high level, works by associating to each constraint a non-negative weight that represents how violated the constraint is. Each iteration of the method uses the current set of weights to decide which variables to update, and increases/decreases the weights by multiplicative factors to reflect the change. This method has been applied, for e.g., to linear, semidefinite and convex programming (e.g., [2,13,18–21]), game theory (e.g. [11]), and geometry (e.g. [6]). Recently, multiplicative weight update methods have been reframed as a specific instance of a more general method known as *mirror descent* [8,25], which is a first-order iterative method for minimizing a convex function over a convex, compact set. This approach is inspired by traditional gradient descent methods, and achieves poly($\frac{1}{\varepsilon}$) oracle complexity. Mirror descent is motivated by the need to maintain almost-dimension-free oracle complexities in situations where bounds on the subgradients of the function being minimized are with respect to arbitrary norms; this is in contrast to gradient descent, which achieves this guarantee only with Euclidean norm bounds. For example, standard implementations of multiplicative weight updates on positive LPs can be derived from mirror descent using l_∞-norm bounds on the subgradients over the simplex. Recently, an algorithm that involves a linear coupling of gradient and mirror descent has been used to derive accelerated methods that perform similarly to Nesterov's accelerated gradient descent for smooth convex functions (i.e., whose gradient is Lipschitz) [26]. A variant of this approach was used to design $\tilde{O}(1/\epsilon)$, width-independent algorithms for packing and covering LPs [29,30].

1.2   Our Contribution and Comparison with Previous Work

We show how to extend the MWU method to solve (PACKING) and (COVERING).
Our algorithm can be thought of as an extension of the work by Garg and
Könemann [13] to handle infinitely-many constraints. The novelty of our approach applied to this problem compared to the finite-constraint case lies in two
key points: (a) In the finite-constraint case, it is straightforward to deduce that
if the potential function (a sum of non-negative components) is bounded, then
each component is bounded by the same amount. In the infinite-constraint case,
deducing a bound on the integrand given a bound on the potential function (an
integral) is more involved (Lemma 4). (b) Bounding the number of iterations of
the algorithm relies on arguments tracking what portion of a given constraint
set has satisfied the corresponding constraint in any given iteration (Lemma 5).

Our approach has the following features: (I) **Oracle-based**: Let $m :=$
$\max_i m_i$, $U := \max_{i,j} a_{ij}$, and assume that each set $\mathcal{Y}_i$ contains a ball of
radius $r_i > 0$ and is contained in a ball of radius R_i. In particular, the sets
$\mathcal{Y}_i$ are *full-dimensional* in their respective spaces. In what follows, we will
denote $R := \max\{R_i, \frac{1}{r_i} : i \in [N]\}$. For the packing (resp., covering) problem, the algorithm is guaranteed to terminate with an ϵ-approximate solution
in[1] $\tilde{O}(\frac{m^2 N}{\epsilon^2} \log(R+1))$ (resp., $\tilde{O}(\frac{m^3 N}{\epsilon^2}) \log^2(RU z^* +1))$ iterations, each requiring
calls to oracle MAXVEC (resp., MINVEC) that computes the argument of the
maximum (resp., minimum) over the sets $\mathcal{Y}_i$, the oracle INTEGRAL that computes
integrals over $\mathcal{Y}_i$ with respect to a log-concave function, and an oracle MINCOL
(resp., MAXCOL) to compute the minimum (resp., maximum) component of the
vectors $w^T A_i$ for a given non negative vector w. Thus, the running time of our
algorithms are only *linear* in the dimension n, and hence become attractive when
m is much smaller than n. (II) **Dimension-independent**: We do not need the
matrices A_i to be explicitly given, as long as the oracle MAXVEC (resp., MIN-
VEC) can be implemented efficiently. In particular, our algorithms are superior
to known methods in the case when there is an exponential number of variables
(see Sect. 2.2 for an application). (III) **Width-independent**: Our algorithms
provide *relative* approximation guarantees, and the running time depends only
polylogarithmically on R, U, and z^*. (IV) **Practical**: For the cases of SOCP, and
robust packing and covering linear programs with simple uncertainty sets such as
balls, the maximization/minimization oracles MAXVEC/MINVEC as well as the
integration oracle INTEGRAL are very easy to compute and hence each iteration
can be implemented to run very fast. We observe furthermore experimentally
that our bounds on the number of iterations are very conservative, and typically
the number of iterations to reach the required accuracy can be much smaller.
This yields substantial improvements in running time over the state-of-the-art
SOCP solvers.

The only algorithms we are aware of for solving general SOCP with theoretical guarantees on the running time are via interior-point methods and take
$O(n^{4.5} \log \epsilon^{-1})$ time to compute an ϵ-approximate solution [15,16]. Note that

---

[1] $\tilde{O}(\cdot)$ suppresses polylogarithmic factors that depend on m, N, and $\frac{1}{\epsilon}$.

even though there are recent improvements via matrix MWU algorithms for solving *positive* SDPs [2,19,30], those techniques cannot be (directly) applied to SOCP since the reduction may lead to non-positive SDPs.

We also note here that recent techniques utilizing the coupling of gradient and mirror descent for solving packing and covering LPs [30] do not seem to extend in a straightforward manner to the case when there are (uncountably) infinitely many constraints. In addition, those methods generally assume the availability of a sampling oracle that can pick uniformly at random from the set of variables (as opposed to an optimization oracle as described in (I) above), which can be a restrictive assumption for some applications. Adapting those techniques to such problems is a potential research direction.

For robust optimization, an oracle-based algorithm was given in [3] which finds an ϵ-approximate solution after $O(F(DG+F)/\epsilon^2)$ calls to an approximate optimization oracle similar to MINCOL/MAXCOL, where $F := \|b\|_2$, G is an upper bound on $\|A_i x\|_2$ over all feasible x, and D is the maximum diameter of the sets $\mathcal{Y}_i$. However, since this algorithm is designed to work for a more general class than (PACKING) and (COVERING), it is not width-independent. It is also worth noting that an extension of the MWU for special convex programs was given in [20]. However, this extension essentially reduces the problem with many convex constraints into one with a single convex constraint and hence is not useful to solve (PACKING) and (COVERING) (except for reducing N to 1).

The rest of this paper is organized as follows. In the next section we briefly explain the application of (PACKING) and (COVERING) to robust optimization and SOCP. In Sect. 3 we describe the packing algorithm and analyze its performance. In Sect. 4, we describe briefly the covering algorithm. Most proofs are omitted due to lack of space.

2 Applications

2.1 Robust Optimization

Consider the following linear programming problem:

$$\min \quad c^T x \quad \text{s.t.} \quad (a^i)^T x \le b_i, \forall i \in [N] \quad x \ge 0, \tag{1}$$

where $c, a^i, b_i \ge 0$. Assume that the first m_i components of the row vector a^i are drawn from an *uncertainty set* $\mathcal{Y}_i \subseteq \mathbb{R}_+^{m_i}$. It is required to solve the LP for the *worst-case* choice of row vectors. We can rewrite the above constraint in (1) as

$$y^T \left[I_{m_i} | 0 \right] x + \left[0 | (\bar{a}^i)^T \right] x \le b_i, \forall y \in \mathcal{Y}_i, \ i \in [N],$$

where I_{m_i} is the $m_i \times m_i$-identity matrix, and $\bar{a}^i$ is the vector consisting of the last $n - m_i$ components of a^i. Similarly, we can consider the covering version of (1).

Typical examples of uncertainty sets include:

– Ellipsoidal uncertainty: $\mathcal{Y}_i = \mathbf{E}(z^i, D_i) := \{y \in \mathbb{R}^{m_i} : (y - z^i)^T D_i^{-2}(y - z^i) \le 1\}$, for given positive definite matrices $D_i \in \mathbb{R}^{m_i \times m_i}$ and vectors $z^i \in \mathbb{R}^{m_i}$.

- Box uncertainty: $\mathcal{Y}_i = \mathbf{B}(z^i, \rho_i) := \{y \in \mathbb{R}^{m_i} : \|y - z^i\|_1 \le \rho_i\}$, for given vectors $z^i \in \mathbb{R}^{m_i}$ and real numbers $\rho_i > 0$.
- Polyhedral uncertainty: $\mathcal{Y}_i := \{y \in \mathbb{R}^{m_i} : D_i y \le w_i\}$, for some $D_i \in \mathbb{R}^{r_i \times m_i}$ and $w_i \in \mathbb{R}^{r_i}$.

When the m_i's are small compared to n and the uncertainty set is simple, e.g., a ball, the integral oracle can be implemented efficiently, yielding significantly faster algorithms than general convex programming solvers.

2.2   Second-Order Cone Programming

Consider a second-order cone programming problem:

$$\max_x c^T x \text{ s.t. } \|A_i x\| + (d^i)^T x \le b_i, \forall i \in [N] \quad x \ge 0, \qquad \text{(SOCP)}$$

where $c, d_i \in \mathbb{R}^n_+$, and $A_i \in \mathbb{R}^{m_i \times n}_+$. The constraint can be rewritten as: $\max_{y \in \mathcal{Y}_i} y^T A_i x + d_i^T x \le b_i, \forall i \in [N]$, where $\mathcal{Y}_i := \{y \in \mathbb{R}^{m_i}_+ : \|y\| \le 1\}$, or $y^T A_i x + d_i^T x \le b_i, \forall y \in \mathcal{Y}_i, i \in [N]$. This formulates (SOCP) as a special case of (PACKING), and implies that we can get much faster (approximation) algorithms, compared to general SOCP solvers, when m is small compared to n.

3   The Packing Problem

We say that $x \in \mathbb{R}^n_+$ is an ϵ-approximate solution for (PACKING) if $c^T x \ge (1 - \epsilon)z^*$. For simplicity of presentation, we assume without loss of generality that $b = \mathbf{1}$ and $c = \mathbf{1}$, where $\mathbf{1}$ is the vector of all ones of appropriate dimension.

3.1   Required Oracles

We assume the availability of the following *oracles*:

- MINCOL(w, ϵ): Given $w \in \mathbb{R}^n_+$ and $\epsilon > 0$, find $j \in [n]$ such that $w^T \mathbf{1}_j \le (1 + \epsilon) \min_{j' \in [n]} w^T \mathbf{1}_{j'}$.
- MAXVEC$(w, \epsilon, \mathcal{Y})$: Given a closed, bounded and convex $\mathcal{Y} \subseteq \mathbb{R}^m_+$, $w \in \mathbb{R}^m_+$ and $\epsilon > 0$, find $y \in \mathcal{Y}$ such that $w^T y \ge (1 - \epsilon) \max_{y' \in \mathcal{Y}} w^T y'$.
- INTEGRAL$(p, f, \epsilon, \sigma, \mathcal{Y})$: Given a closed, bounded and convex $\mathcal{Y} \subseteq \mathbb{R}^m_+$, a log-concave[2] function $p : Y \to \mathbb{R}_+$, a function $f : Y \to \mathbb{R}^k$, and $\epsilon, \sigma \in [0, 1)$, find $\bar{f} \in \mathbb{R}^k$ such that

$$\Pr\left[\left|\bar{f} - \int_Y p(y)f(y)dy\right| \le \epsilon \cdot \int_Y p(y)f(y)dy\right] \ge 1 - \sigma.$$

We will use INTEGRAL$(\cdots)$ only with $f(y) := 1$ or $f(y) := y$ which implies that (each component of) $f(y)p(y)$ is also log-concave whenever $p(y)$ is.

---

[2] That is, $\log p$ is concave.

3.2   Algorithm for Packing

The algorithm is shown in Algorithm 1. For $i \in [N]$ and $y \in \mathcal{Y}_i$, we define $g_i(y) := A_i^T y + d^i$. Recall $m := \max_{i \in [N]} m_i$. At a high level, the algorithm maintains a set of weights $p_i(y, t) := (1 + \epsilon)^{g_i(y)^T x(t)}$, for $i \in [N]$ and $y \in \mathcal{Y}_i$. In each iteration t, it computes (approximately) the integral $w := \sum_i \int_{\mathcal{Y}_i} p_i(y, t) g_i(y) dy$ and calls the MinCol oracle to find a column j minimizing $w^T \mathbf{1}_j$. Then the j'th component of x is increased by $\delta(t) \approx 1/\max_i \max_{y \in \mathcal{Y}_i} g_i(y)^T \mathbf{1}_j$. The algorithm iterates until $M(t) \approx \max_i \max_{y \in \mathcal{Y}_i} g_i(y)^T x(t)$ becomes sufficiently large; more precisely, equal to some parameter T which is fixed appropriately.

3.3   Analysis of Packing Algorithm

Towards showing that the algorithm terminates with an ϵ-approximate solution for (Packing), we define the following *potential function*: $\Phi(t) := \sum_i \Phi_i(t)$ where $\Phi_i(t) := \int_{\mathcal{Y}_i} p_i(y, t) dy$.

Outline of the Analysis. We analyze the algorithm in three parts. In the first part, we bound the potential increase from one iteration to the next (Lemma 1), and use this to relate the potential after t iterations to the initial potential and the ratio of the value of the current solution to z^* (Lemmas 2 and 3). However, bounding $\Phi(t)$ does not directly imply that we arrive at an ϵ-approximate solution, due to the fact that a definite integral of a non-negative function over a given convex region $\mathcal{Y}_i$ being bounded by some γ does not imply that the function at any point in $\mathcal{Y}_i$ is also bounded by γ. In the second part (Lemma 4), we overcome this difficulty by showing that, due to the convexity of the sets $\mathcal{Y}_i$, the value at a given point cannot be large unless there is a sufficiently large fraction of the volume of the set $\mathcal{Y}_i$ over which the integral is also large. (This extends a lemma proved by some of the authors in previous work [10]). We use this to show in Sect. 3.5 that the algorithm converges to an ϵ-approximate solution, assuming the existence of suitable parameters satisfying certain conditions, which are shown to exist in Sect. 3.6. In the third part of the analysis, we bound the number of iterations by showing that the algorithm makes sufficient progress in each iteration, by satisfying constraints corresponding to a large fraction of the volume of one of the sets $\mathcal{Y}_i$.

The following three lemmas are obtained by standard analysis of MWU methods (see, e.g., [13]) with "$\sum$"'s replaced by "$\int$"'s.

Lemma 1. $\Phi(t + 1) \leq \Phi(t) \exp\left(\epsilon \delta(t) \sum_i \int_{\mathcal{Y}_i} \frac{p_i(y,t)}{\Phi(t)} g_i(y)^T \mathbf{1}_{j(t)} dy\right).$

Define $\bar{\epsilon} := \frac{(1+\epsilon_2)(1+\epsilon_1)}{1-\epsilon_1}$.

Lemma 2. Let $\kappa(t) := \sum_{t'=0}^{t-1} \delta(t') \sum_i \int_{\mathcal{Y}_i} \frac{p_i(y,t')}{\Phi(t')} g_i(y)^T \mathbf{1}_{j(t')} dy$. Then with probability at least $1 - 2N\sigma t$, $\kappa(t) \leq \frac{(1+\bar{\epsilon}) \mathbf{1}^T x(t)}{z^*}$ for all t.

Data: Matrices $A_i \in \mathbb{R}_+^{m_i \times n}$, $d^i \in \mathbb{R}_+^n$, sets $\mathcal{Y}_i \subseteq \mathbb{R}^{m_i}$, for $i \in [N]$, and an approximation accuracy $\epsilon \in (0, 1)$, probability $\sigma \in (0, 1)$.

Result: An $O(\epsilon)$-approximate solution $\hat{x}$ for (PACKING).

1   $\epsilon_1 = \epsilon_2 = \epsilon_3 \leftarrow \epsilon$

2   $V \leftarrow \max_i \max\{2\pi^{m/2} R_i^m, \frac{1}{\pi^{m/2} r_i^m/(2m)^m}\}$

3   $\alpha \leftarrow \frac{2}{\log(1+\epsilon)} \left(1 + \log N + m \log \frac{9}{\epsilon^2 (1-\epsilon_3)^2} + \log V\right)$

4   $T \leftarrow \frac{1}{\epsilon^2} (\alpha \log(1 + \epsilon) + \log N + \log V)$

5   $t \leftarrow 0; x(0) \leftarrow 0, \forall i \in [N]; M(0) \leftarrow 0$

6   **while** $M(t) < T$ **do**

7   **for** $i \in [N]$ **do**

8          $p_i(y, t) \leftarrow (1 + \epsilon)^{g_i(y)^T x(t)}$

9          $\bar{y}^i(t) \leftarrow \text{INTEGRAL}(p_i(y, t), f(y) := y, \epsilon_1, \sigma, \mathcal{Y}_i)$

10         $\bar{\phi}^i(t) \leftarrow \text{INTEGRAL}(p_i(y, t), f(y) := 1, \epsilon_1, \sigma, \mathcal{Y}_i)$

11      $j(t) \leftarrow \text{MINCOL}(\sum_i (A_i^T \bar{y}^i(t) + \bar{\phi}^i(t) d^i), \epsilon_2)$

12   **for** $i \in [N]$ **do**

13          $\hat{y}^i(t) \leftarrow \text{MAXVEC}(A_i \mathbf{1}_{j(t)}, \epsilon_3, \mathcal{Y}_i)$

14      $\delta(t) \leftarrow \frac{1 - \epsilon_3}{\max_i g_i(\hat{y}^i(t))^T \mathbf{1}_{j(t)}}$

15      $x(t + 1) \leftarrow x(t) + \delta(t) \mathbf{1}_{j(t)}$

16   **for** $i \in [N]$ **do**

17          $\tilde{y}^i(t + 1) \leftarrow \text{MAXVEC}(A_i x(t + 1), \epsilon_3, \mathcal{Y}_i)$

18      $M(t + 1) \leftarrow \max_i g_i(\tilde{y}^i(t + 1))^T x(t + 1)$

19      $t \leftarrow t + 1$

20   $\hat{x} = \frac{(1 - \epsilon_3) x(t)}{M(t)}$

Output: $\hat{x}$

Algorithm 1: The packing algorithm

Lemma 3. *For all t, with probability at least $1 - 2N\sigma t$, it holds that*

$$\Phi(t) \le \Phi(0) \exp\left(\frac{\epsilon(1 + \bar{\epsilon})\mathbf{1}^T x(t)}{z^*}\right). \tag{2}$$

The next step is to bound each term under the integral in $\Phi(t)$, given that this sum of integrals is bounded as in (2). For this we slightly modify a lemma in [10].

Lemma 4. *Suppose $\Phi(t) \le \gamma$ for some $\gamma > 0$ and some iteration t of the algorithm. Also, suppose that for given $T > 0$ and $\epsilon_3 > 0$ there exists an α, such that*

$$0 < \alpha < \alpha_0, \tag{3}$$

$$(1 + \epsilon)^{\alpha/2} \min_i \left\{\left(\frac{\alpha}{\alpha_0}\right)^{m_i} \Phi_i(0)\right\} > 1, \tag{4}$$

where $\alpha_0 := 2\frac{T+1-\epsilon_3}{(1-\epsilon_3)^2}$. Then $(1 + \epsilon)^{g_i(y)^T x(t)} \le \gamma(1 + \epsilon)^\alpha$ for all $y \in \mathcal{Y}_i$, for all $i \in [N]$.

3.4   Bounding the Number of Iterations

Lemma 5. *With probability at least $1 - 2N\sigma t_f$, Algorithm 1 terminates in $t_f \leq \frac{2NT}{(1-\epsilon_3)^2(1-\lambda)}$ iterations, where σ is the probability parameter used in the oracle* INTEGRAL$(\cdots)$.

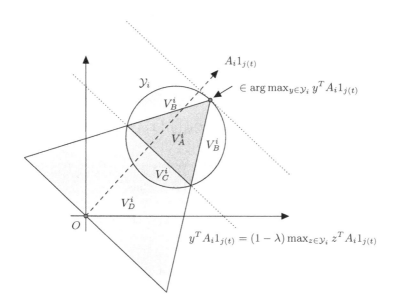

Fig. 1. Illustration of Lemma 5

Proof. Consider iteration t of the algorithm. Let $H_i(t, \lambda)$ be the hyperplane $\{y \in \mathbb{R}^{m_i} : y^T A_i \mathbf{1}_{j(t)} = (1 - \lambda) \max_{z \in \mathcal{Y}_i} z^T A_i \mathbf{1}_{j(t)}\}$. Define $C_i(t)$ to be the cone whose apex is (one of the points in) $\arg\max_{z \in \mathcal{Y}_i} z^T A_i \mathbf{1}_{j(t)}$, and one of whose cross-sections is $H_i(t, \lambda) \cap \mathcal{Y}_i$; i.e., the portion of the hyperplane lying in $\mathcal{Y}_i$. Finally, let $\hat{H}_i(t, \lambda) = \{y \in \mathcal{Y}_i : y^T A_i \mathbf{1}_{j(t)} \geq (1 - \lambda) \max_{z \in \mathcal{Y}_i} z^T A_i \mathbf{1}_{j(t)}\}$.

We notate the following volumes (illustrated in Fig. 1) as:

$$V_A^i = \text{vol}(C_i(t) \cap \hat{H}_i(t, \lambda)); \quad V_B^i = \text{vol}(\mathcal{Y}_i \setminus C_i(t));$$
$$V_C^i = \text{vol}(\mathcal{Y}_i \setminus \hat{H}_i(t, \lambda)); \quad V_D^i = \text{vol}(C_i(t) \cap \hat{H}_i(t, 1) \setminus \mathcal{Y}_i).$$

By similarity, $V_A^i = \lambda^{m_i}(V_A^i + V_C^i + V_D^i)$. Thus,

$$V_A^i = \frac{\lambda^{m_i}}{1 - \lambda^{m_i}}(V_C^i + V_D^i).$$

Now,

$$V_A^i + V_B^i \geq V_A^i = \frac{\lambda^{m_i}}{1 - \lambda^{m_i}}(V_C^i + V_D^i) \geq \frac{\lambda^{m_i}}{1 - \lambda^{m_i}}V_C^i.$$

Therefore $(1 - \lambda^{m_i})(V_A^i + V_B^i) \geq \lambda^{m_i} V_C^i$. Adding $\lambda^{m_i}(V_A^i + V_B^i)$ to both sides, we have:

$$V_A^i + V_B^i \geq \lambda^{m_i}(V_A^i + V_B^i + V_C^i) = \left(1 - \frac{1}{e}\right)^{\frac{m_i}{2m}}(V_A^i + V_B^i + V_C^i).$$

For $m_i \geq 1$, $\left(1 - \frac{1}{e}\right)^{\frac{m_i}{2m}} \geq \frac{1}{2}$. Thus, for all i and t,

$$\text{vol}(\hat{H}_i(t,\lambda)) \geq \frac{1}{2}\text{vol}(\mathcal{Y}_i). \tag{5}$$

Fix $i(t) \in \arg\max_i \max_{z \in \mathcal{Y}_i} g_i(z)^T \mathbf{1}_{j(t)}$ for each $t = 0, 1, 2 \ldots$ Note that for any i,

$$g_i(y)^T \Delta x(t) := g_i(y)^T [x(t+1) - x(t)] = \delta(t) g_i(y)^T \mathbf{1}_{j(t)} \geq \frac{(1 - \epsilon_3)g_i(y)^T \mathbf{1}_{j(t)}}{\max_{z \in \mathcal{Y}_{i(t)}} g_{i(t)}(z)^T \mathbf{1}_{j(t)}}.$$

If $y \in \hat{H}_{i(t)}(t,\lambda)$ then $y^T A_{i(t)} \mathbf{1}_{j(t)} \geq (1 - \lambda) \max_{z \in \mathcal{Y}_{i(t)}} z^T A_{i(t)} \mathbf{1}_{j(t)}$, and hence, $g_{i(t)}(y)^T \mathbf{1}_{j(t)} \geq (1 - \lambda) \max_{z \in \mathcal{Y}_{i(t)}} g_{i(t)}(z)^T \mathbf{1}_{j(t)}$. Thus,

$$y \in \hat{H}_{i(t)}(t,\lambda) \implies g_{i(t)}(y)^T \Delta x(t) \geq (1 - \epsilon_3)(1 - \lambda). \tag{6}$$

Suppose for the sake of contradiction that $t_f > t_0 := \frac{2NT}{(1-\epsilon_3)^2(1-\lambda)}$. Then $M(t_0) < T$, implying that $g_i(y)^T x(t_0) < \frac{T}{1-\epsilon_3}$ for all i and $y \in \mathcal{Y}_i$. For $t' = 0, 1, \ldots, t_0$, and $y \in \cup_i \mathcal{Y}_i$, define an indicator variable $I(t', y)$ that takes value 1 if and only if $y \in \hat{H}_{i(t')}(t',\lambda)$. By (6), if $y \in \mathcal{Y}_i$ and $r(y) := \sum_{t'} I(t', y) = |\{t' \in \{0, 1, \ldots, t_0\} : y \in \hat{H}_{i(t')}(t',\lambda)\}|$, then $g_i(y)^T x(t_0) > (1 - \epsilon_3)(1 - \lambda)r(y)$, and hence, $r(y) < \frac{T}{(1-\epsilon_3)^2(1-\lambda)}$. It follows that

$$\frac{T}{(1-\epsilon_3)^2(1-\lambda)}\text{vol}(\mathcal{Y}_i) = \int_{\mathcal{Y}_i} \frac{T}{(1-\epsilon_3)^2(1-\lambda)}dy > \int_{\mathcal{Y}_i} r(y)dy$$

$$= \int_{\mathcal{Y}_i} \sum_{t'} I(t', y)dy = \sum_{t'} \int_{\mathcal{Y}_i} I(t', y)dy$$

$$= \sum_{t': i(t')=i} \int_{\hat{H}_i(t',\lambda)} dy = \sum_{t': i(t')=i} \text{vol}(\hat{H}_{i(t')}(t',\lambda))$$

$$\geq \sum_{t': i(t')=i} \frac{\text{vol}(\mathcal{Y}_i)}{2} = \frac{t_i}{2}\text{vol}(\mathcal{Y}_i),$$

where $t_i := |\{t' : i(t') = i\}|$. Thus, we obtain $t_0 = \sum_i t_i < \frac{2NT}{(1-\epsilon_3)^2(1-\lambda)}$, a contradiction. Thus, the algorithm terminates in $\frac{2NT}{(1-\epsilon_3)^2(1-\lambda)}$ iterations.

3.5   Convergence to an ϵ-Approximate Solution

Lemma 6. *Suppose (2) holds and T and α satisfy the hypothesis of Lemma 4. If furthermore $\epsilon \leq 1$ and* [3]

$$T \geq \frac{\log \Phi(0) + \alpha \log(1+\epsilon)}{\epsilon^2} \qquad (7)$$

then Algorithm 1 terminates with a $\frac{(1-\epsilon_3)(1-2\epsilon)}{(1+\bar{\epsilon})}$-approximate solution $\hat{x}(t)$ for (PACKING).

Proof. We set γ in Lemma 4 to $\Phi(0) \exp\left(\frac{\epsilon(1+\bar{\epsilon})\mathbf{1}^T x(t)}{z^*}\right)$ to conclude that, for all $y \in \mathcal{Y}_i$, all $i \in [N]$, and all t,

$$(1+\epsilon)^{g_i(y)^T x(t)} \leq \Phi(0) \exp\left(\frac{\epsilon(1+\bar{\epsilon})\mathbf{1}^T x(t)}{z^*}\right)(1+\epsilon)^\alpha.$$

$$\therefore g_i(y)^T x(t) \log(1+\epsilon) \leq \log \Phi(0) + \alpha \log(1+\epsilon) + \frac{\epsilon(1+\bar{\epsilon})\mathbf{1}^T x(t)}{z^*}.$$

Dividing by $\epsilon(1+\bar{\epsilon})M(t)$ and rearranging,

$$\frac{\mathbf{1}^T x(t)}{z^* M(t)} \geq \frac{\log(1+\epsilon)g_i(y)^T x(t)}{\epsilon(1+\bar{\epsilon})M(t)} - \frac{\log \Phi(0) + \alpha \log(1+\epsilon)}{\epsilon(1+\bar{\epsilon})M(t)}. \qquad (8)$$

Define $i(t) \in [N]$ and $y(t) \in \mathcal{Y}_{i(t)}$ such that $M(t)$ is set to $g_{i(t)}(y(t))^T x(t)$ in step 18 of the algorithm. Using these particular i and y in (8), we get

$$\frac{\mathbf{1}^T x(t)}{z^* M(t)} \geq \frac{\log(1+\epsilon)}{\epsilon(1+\bar{\epsilon})} - \frac{\log \Phi(0) + \alpha \log(1+\epsilon)}{\epsilon(1+\bar{\epsilon})M(t)}.$$

Suppose the algorithm terminates after t_f iterations. Then $M(t_f) \geq T$ (for T predefined in step 4 of the algorithm). Then

$$\frac{\mathbf{1}^T x(t_f)}{z^* M(t_f)} \geq \frac{\log(1+\epsilon)}{\epsilon(1+\bar{\epsilon})} - \frac{\log \Phi(0) + \alpha \log(1+\epsilon)}{\epsilon(1+\bar{\epsilon})M(t_f)}$$

$$\geq \frac{\log(1+\epsilon)}{\epsilon(1+\bar{\epsilon})} - \frac{\log \Phi(0) + \alpha \log(1+\epsilon)}{\epsilon(1+\bar{\epsilon})T}.$$

Since $T \geq \frac{\log \Phi(0) + \alpha \log(1+\epsilon)}{\epsilon^2}$, we have:

$$\frac{\mathbf{1}^T x(t_f)}{z^* M(t_f)} \geq \frac{1}{1+\bar{\epsilon}}\left(\frac{\log(1+\epsilon)}{\epsilon} - \epsilon\right) \geq \frac{1-2\epsilon}{1+\bar{\epsilon}}$$

for $\epsilon \leq 1$. Thus, at the termination of the algorithm, the output $\hat{x}(t) = \frac{(1-\epsilon_3)x(t_f)}{M(t_f)}$ is feasible by the definition of $M(t_f)$ and achieves the required result.

---

[3] Throughout, "log" denotes the *natural* logarithm.

3.6   Existence of Suitable Parameters

Lemma 7. *For any $\epsilon \in (0,1)$, there exist α and*

$$T = O\left(\frac{\log N + m \log \frac{Rm}{\epsilon}}{\epsilon^2}\right)$$

satisfying (3), (4) and (7).

Finally note that, according to Lemma 2, the bound of Lemma 5 on the running time t_f holds with probability at least $1 - 2N\sigma t_f$. Selecting σ small enough, e.g., $\sigma = \frac{1}{2mNt_f}$, gives us the final result for (PACKING).

Theorem 1. *For any $\epsilon > 0$, there is an algorithm that computes, with probability $1 - o(1)$, an ϵ-approximate solution for (PACKING) using $\tilde{O}(\frac{m^2 N}{\epsilon^2} \log(R+1))$ calls to the oracles* MINCOL$(\cdots)$, INTEGRAL$(\cdots)$ *and* MAXVEC$(\cdots)$.

4   The Covering Problem

We say that $x \in \mathbb{R}^n_+$ is an ϵ-approximate solution for (COVERING) if $c^T x \leq (1+\epsilon)z^*$.

In addition to the oracles INTEGRAL$(\cdots)$ and MAXVEC$(\cdots)$, we assume the availability of the following *oracles*:

- MAXCOL(w, ϵ): Given $w \in \mathbb{R}^n_+$ and $\epsilon > 0$, find $j \in [n]$ such that $w^T \mathbf{1}_j \geq (1 - \epsilon) \max_{j' \in [n]} w^T \mathbf{1}_{j'}$.
- MINVEC$(w, \epsilon, \mathcal{Y})$: Given a closed, bounded and convex $\mathcal{Y} \subseteq \mathbb{R}^m_+$, $w \in \mathbb{R}^m_+$ and $\epsilon > 0$, find $y \in \mathcal{Y}$ such that $w^T y \leq (1 + \epsilon) \min_{y' \in \mathcal{Y}} w^T y'$.

The algorithm for (COVERING) proceeds largely analogously to the case of packing; see Algorithm 2. However, the analysis is much more complicated due to the fact that we have to truncate the sets $\mathcal{Y}_i$ into their *active* subsets $\mathcal{Y}_i(t)$ in each iteration, defined as $\mathcal{Y}_i(t) := \{y \in \mathcal{Y}_i : g_i(y)^T x(t) \leq T\}$. As the volumes of these sets get smaller, the bounds in the lemma analogous to Lemma 4 will not apply, and hence, we have also to maintain a list of active sets $I(t)$ whose volumes are sufficiently large. Bounding the running time is also more tricky, and we manage to do it by tracking the *centroids* of the sets $\mathcal{Y}_i(t)$. We state here only the main theorem:

Theorem 2. *For any $\epsilon > 0$, there is an algorithm that computes, with probability $1 - o(1)$, an ϵ-approximate solution for (COVERING) using $\tilde{O}(\frac{m^3 N}{\epsilon^2} \log^2(RUz^* + 1) \log(z^*))$ calls to the oracles* MINCOL$(\cdots)$, INTEGRAL$(\cdots)$, MINVEC$(\cdots)$ *and* MAXVEC$(\cdots)$, *where $U := \max_{i,j} a_{ij}$.*

Data: Matrices $A_i \in \mathbb{R}_+^{m_i \times n}$, $d^i \in \mathbb{R}_+^n$, sets $\mathcal{Y}_i \subseteq \mathbb{R}^{m_i}$, for $i \in [N]$, and an approximation accuracy $\epsilon \in (0,1)$, probability $\sigma \in (0,1)$.

Result: An $O(\epsilon)$-approximate solution $\hat{x}$ for (COVERING).

1   $\lambda \leftarrow (1 - 1/e)^{\frac{1}{2m}}$; $\epsilon_1 = \epsilon_2 = \epsilon_3 = \epsilon_4 \leftarrow \epsilon$; $\theta \leftarrow \lambda^m$

2   $\bar{z} \leftarrow 1$

3   $V \leftarrow \max_i \max\{2\pi^{m/2} R_i^m, \frac{1}{\pi^{m/2} r_i^m/(2m)^m}\}$

4   **repeat**

5      $\bar{U} := R(\sqrt{m} + 1)U\frac{\bar{z}}{(1-\bar{\epsilon})}$

6      $a \leftarrow \max\{1, m/\log(1/\theta)\}$

7      $b \leftarrow \max\{1, \log \frac{2\bar{U}N}{\epsilon_4(1-\epsilon_3)(1-\lambda)}\}$

8      $\bar{\ell} = 5ab(\log(a + e - 1) + 1)$

9      $\alpha \leftarrow \frac{2}{\log(1-\epsilon)^{-1}}\left(1 + \log N + m\log \frac{3}{\epsilon^2} + \bar{\ell}\log\frac{1}{\theta} + \log V\right)$

10     $T \leftarrow \frac{1}{\epsilon^2}\left(1 + \alpha \log(1-\epsilon)^{-1} + \log N + \bar{\ell}\log\frac{1}{\theta} + \log V\right)$

11     $t \leftarrow 0$; $x(0) \leftarrow 0$, $\forall i \in [N]$; $I(0) \leftarrow [N]$

12   **while** $I(t) \neq \emptyset$ **do**

13   **for** $i \in I(t)$ **do**

14           $p_i(y, t) \leftarrow (1 - \epsilon)^{g_i(y)^T x(t)}$

15           $\bar{y}^i(t) \leftarrow$ INTEGRAL$(p_i(y, t), f(y) := y, \epsilon_1, \sigma, \mathcal{Y}_i(t))$

16           $\bar{\phi}^i(t) \leftarrow$ INTEGRAL$(p_i(y, t), f(y) := 1, \epsilon_1, \sigma, \mathcal{Y}_i(t))$

17        $j(t) \leftarrow$ MAXCOL$(\sum_{i \in I(t)}(A_i^T \bar{y}^i + \bar{\phi}^i(t)d^i), \epsilon_2)$

18   **for** $i \in I(t)$ **do**

19           $\hat{y}^i(t) \leftarrow$ MAXVEC$(A_i \mathbf{1}_{j(t)}, \epsilon_3, \mathcal{Y}_i(t))$

20        $\delta(t) \leftarrow \frac{1-\epsilon_3}{\max_{i \in I(t)} g_i(\hat{y}^i(t))^T \mathbf{1}_{j(t)}}$

21        $x(t+1) \leftarrow x(t) + \delta(t)\mathbf{1}_{j(t)}$

22        $I(t+1) \leftarrow I(t)$

23   **for** $i \in I(t)$ **do**

24           $\tilde{y}^i(t+1) \leftarrow$ MINVEC$(A_i x(t+1), \epsilon_3, \mathcal{Y}_i(t+1))$

25   **if** $g_i(\tilde{y}^i(t+1))^T x(t+1) \geq (1 - \epsilon_4)T$ **then**

26              $I(t+1) \leftarrow I(t+1)\backslash\{i\}$

27        $t \leftarrow t + 1$

28     $\hat{x} = \frac{x(t-1)}{\left(\frac{1-\epsilon_4}{1+\epsilon_3} - \epsilon^2\right)T}$

29     $\bar{z} \leftarrow 2\bar{z}$

30 **until** $t < \left(\frac{T}{(1-\epsilon_3)(1-\lambda)} + 1\right) N\bar{\ell}$;

Output: $\hat{x}$

Algorithm 2: The covering algorithm

References

1. Alizadeh, F., Goldfarb, D.: Second-order cone programming. Math. Program. **95**(1), 3–51 (2003)
2. Arora, S., Hazan, E., Kale, S.: Fast algorithms for approximate semidefinite programming using the multiplicative weights update method. In: FOCS, pp. 339–348 (2005)

3. Ben-Tal, A., Hazan, E., Koren, T., Mannor, S.: Oracle-based robust optimization via online learning. Oper. Res. **63**(3), 628–638 (2015)
4. Ben-Tal, A., Nemirovski, A.: Robust optimization - methodology and applications. Math. Program. **92**(3), 453–480 (2002)
5. Bertsimas, D., Thiele, A.: A robust optimization approach to supply chain management. In: Bienstock, D., Nemhauser, G. (eds.) IPCO 2004. LNCS, vol. 3064, pp. 86–100. Springer, Heidelberg (2004). doi:10.1007/978-3-540-25960-2_7
6. Brönnimann, H., Goodrich, M.T.: Almost optimal set covers in finite VC-dimension. Discrete Comput. Geom. **14**(4), 463–479 (1995)
7. Brown, G.W.: Iterative solution of games by fictitious play. In: Koopmans, T.C. (ed.) Activity Analysis of Production and Allocation, pp. 374–376 (1951)
8. Bubeck, S.: Convex optimization: algorithms and complexity. Found. Trends Mach. Learn. **8**(3–4), 231–357 (2015)
9. Chau, C.-K., Elbassioni, K., Khonji, M.: Truthful mechanisms for combinatorial AC electric power allocation. In: AAMAS, pp. 1005–1012 (2014)
10. Elbassioni, K., Makino, K., Mehlhorn, K., Ramezani, F.: On randomized fictitious play for approximating saddle points over convex sets. Algorithmica **73**(2), 441–459 (2015)
11. Freund, Y., Schapire, R.E.: Adaptive game playing using multiplicative weights. Games Econ. Behav. **29**(1–2), 79–103 (1999)
12. Garg, N., Khandekar, R.: Fractional covering with upper bounds on the variables: solving LPs with negative entries. In: Albers, S., Radzik, T. (eds.) ESA 2004. LNCS, vol. 3221, pp. 371–382. Springer, Heidelberg (2004). doi:10.1007/978-3-540-30140-0_34
13. Garg, N., Könemann, J.: Faster and simpler algorithms for multicommodity flow and other fractional packing problems. In: Proceedings 39th Symposium Foundations of Computer Science (FOCS), pp. 300–309 (1998)
14. Goldfarb, D., Iyengar, G.: Robust portfolio selection problems. Math. Oper. Res. **28**(1), 1–38 (2003)
15. Grant, M., Boyd, S.: Graph implementations for nonsmooth convex programs. In: Blondel, V., Boyd, S., Kimura, H. (eds.) Recent Advances in Learning and Control. LNCIS, vol. 371, pp. 95–110. Springer, Heidelberg (2008). doi:10.1007/978-1-84800-155-8_7
16. Grant, M., Boyd, S.: CVX: Matlab software for disciplined convex programming, version 2.1, March 2014
17. Grigoriadis, M.D., Khachiyan, L.G.: Approximate solution of matrix games in parallel. In: Advances in Optimization and Parallel Computing, pp. 129–136 (1992)
18. Grigoriadis, M.D., Khachiyan, L.G.: A sublinear-time randomized approximation algorithm for matrix games. Oper. Res. Lett. **18**(2), 53–58 (1995)
19. Jain, R., Yao, P.: A parallel approximation algorithm for positive semidefinite programming. In: FOCS, pp. 463–471 (2011)
20. Khandekar, R.: Lagrangian relaxation based algorithms for convex programming problems. Ph.D. thesis, Indian Institute of Technology, Delhi (2004)
21. Koufogiannakis, C., Young, N.E.: Beating simplex for fractional packing and covering linear programs. In: FOCS, pp. 494–504 (2007)
22. Littlestone, N., Warmuth, M.K.: The weighted majority algorithm. Inf. Comput. **108**(2), 212–261 (1994)
23. López, M., Still, G.: Semi-infinite programming. Eur. J. Oper. Res. **180**(2), 491–518 (2007)
24. Lovász, L., Vempala, S.: Fast algorithms for logconcave functions: sampling, rounding, integration and optimization. In: FOCS, pp. 57–68 (2006)

25. Nemirovski, A.S., Yudin, D.B.: Problem Complexity and Method Efficiency in Optimization. A Wiley-Interscience Publication. Wiley, Hoboken (1983)
26. Nesterov, Y.: A method of solving a convex programming problem with convergence rate O(1/sqr(k)). Sov. Math. Dokl. **27**, 372–376 (1983)
27. Robinson, J.: An iterative method of solving a game. Ann. Math. **54**(2), 296–301 (1951)
28. Shivaswamy, P.K., Bhattacharyya, C., Smola, A.J.: Second order cone programming approaches for handling missing and uncertain data. J. Mach. Learn. Res. **7**, 1283–1314 (2006)
29. Wang, D., Rao, S., Mahoney, M.W.: Unified acceleration method for packing, covering problems via diameter reduction. CoRR abs/1508.02439 (2015). http://arxiv.org/abs/1508.02439. To appear in ICALP 2016
30. Zhu, Z. A., Orecchia, L.: Nearly-linear time positive LP solver with faster convergence rate. In: STOC, pp. 229–236 (2015)

Balanced Optimization with Vector Costs

Annette M.C. Ficker[1]([✉]), Frits C.R. Spieksma[1], and Gerhard J. Woeginger[2]

[1] Operations Research Group, KU Leuven, Leuven, Belgium
{annette.ficker,frits.spieksma}@kuleuven.be
[2] Department of Mathematics, TU Eindhoven, Eindhoven, Netherlands
gwoegi@win.tue.nl

Abstract. An instance of a *balanced optimization problem with vector costs* consists of a ground set X, a vector cost for every element of X, and a system of feasible subsets over X. The goal is to find a feasible subset that minimizes the spread (or imbalance) of values in every coordinate of the underlying vector costs.

We investigate the complexity and approximability of balanced optimization problems in a fairly general setting. We identify a large family of problems that admit a 2-approximation in polynomial time, and we show that for many problems in this family this approximation factor 2 is best-possible (unless $P = NP$). Special attention is paid to the balanced assignment problem with vector costs, which is shown to be NP-hard even in the highly restricted case of sum costs.

Keywords: Balanced optimization · Assignment problem · Computational complexity · Approximation

1 Introduction

Balanced optimization with vector costs is a family of optimization problems, which extends the work of Martello et al. [10]. The details for this framework will be given later, for now we concentrate on one problem in this framework, the balanced assignment problem.

In the balanced assignment problem (Martello et al. [10]), we are given an $n \times n$ matrix C with real entries $c(i,j)$ for $1 \leq i, j \leq n$. An *assignment* A is a set of n matrix entries that contains exactly one entry from every row and every column. The *imbalance* of assignment A is given by

$$\max_{(i,j) \in A} c(i,j) - \min_{(i,j) \in A} c(i,j),$$

and the goal is to find an assignment that minimizes the imbalance. In a generalization of this problem, the entries $c(i,j)$ are not real scalars but real vectors $\mathbf{c}(i,j)$ of length d; that is

$$\mathbf{c}(i,j) = (c_1(i,j), c_2(i,j), \ldots, c_d(i,j)), \quad \text{for } 1 \leq i, j \leq n.$$

© Springer International Publishing AG 2017
K. Jansen and M. Mastrolilli (Eds.): WAOA 2016, LNCS 10138, pp. 92–102, 2017.
DOI: 10.1007/978-3-319-51741-4_8

The imbalance in the k-th coordinate of assignment A (with $1 \le k \le d$) is

$$\Delta_k(A) = \max_{(i,j)\in A} c_k(i,j) - \min_{(i,j)\in A} c_k(i,j),$$

and the imbalance of assignment A is finally given by

$$\Delta_{\max}(A) = \max_k \Delta_k(A).$$

The objective in the *balanced assignment problem with vector costs* is to find an assignment A that minimizes the imbalance $\Delta_{\max}(A)$. Note that for $d = 1$ we recover the traditional balanced assignment problem.

Apart from being a natural generalization of the traditional balanced assignment problem, there are practical applications of balanced optimization problems with vector costs documented in the literature. For instance, Kamura and Nakamori [6] sketch an industrial problem in the manufacturing of glass lenses that gives rise to a (specially structured) balanced assignment problem with vector costs; see Sect. 5 for more details on this.

1.1 Related Literature

Martello et al. [10] introduce a framework containing many balanced optimization problems with scalar costs, and present a polynomial time algorithm to solve these problems. We now discuss some of these problems in more detail.

In the balanced version of the shortest path problem, we are given a directed graph $G = (V, E)$, two nodes s and t, and scalar costs on the edges. The goal is to find a path from s to t that minimizes the difference between the largest and the smallest edge cost along the path. Turner [11] generalizes this problem to finding a path that minimizes the difference between the k_1-th largest and the k_2-th smallest edge cost, and shows that this problem is solvable in polynomial time. Cappanera and Scutellá [3] discuss other balanced path problems. Their goal is to identify p (arc-disjoint or node disjoint) paths from s to t, such that the difference between the length of longest path and the length of the shortest path is minimal. These problems are NP-hard, even for $p = 2$.

In the balanced version of the minimum cut problem, we are given an undirected graph $G = (V, E)$, two nodes s and t, and scalar costs on the edges. The goal is to find a cut that minimizes the difference between the largest and the smallest cost of edges in the cut. Katoh and Iwano [7] construct an algorithm for this problem with running time $O\left(\mathrm{MST}\left(|V|, |E|\right) + |V| \log |V|\right)$, where $\mathrm{MST}\left(|V|, |E|\right)$ denotes the running time for computing the minimum and maximum spanning trees in a graph $G = (V, E)$.

In the balanced version of the spanning tree problem, we are given a graph $G = (V, E)$ and scalar costs on the edges. The goal is to find a spanning tree that minimizes the difference between the largest and the smallest edge cost in the spanning tree. Camerini et al. [2] and Galil and Schieber [5] construct algorithms for this problem, with running times $O(|V| \cdot |E|)$ and $O(|E| \log |V|)$ respectively.

In the balanced version of the traveling salesman problem, we are given a graph $G = (V, E)$ and scalar costs on the edges. The goal is to find a Hamiltonion cycle that minimizes the difference between the largest and the smallest edge cost in cycle. This problem is obviously NP-hard, and Larusic and Punnen [9] discuss several heuristics for it. Kinable et al. [8] discuss a related problem, called the equitable traveling salesman problem. They observe that a Hamiltonian cycle is the union of two uniquely defined matchings. Their goal is to find a Hamiltonian cycle in which the difference between the total cost of its two matchings is minimal.

Another interesting problem in this area is the balanced version of linear programming. Here we are given a system of linear constraints ($Ax = b$ and $x \geq 0$) and costs associated with each real variable x_i. The goal is to minimize the difference between the largest non-zero cost $c_i x_i$ and the smallest non-zero cost $c_j x_j$. Ahuja [1] presents a polynomial time algorithm for this problem.

Finally, an example of an optimization problem featuring vector costs is described by Dokka et al. [4]; we stress however that the objective in the underlying multi-index assignment problem is quite different from minimizing imbalance.

1.2   Our Results

We derive a variety of results on the complexity and approximability of balanced optimization problems with vector costs:

- First, we describe a framework for balanced optimization problems that takes vector costs into account, thereby extending the work of Martello et al. [10]; see Sect. 2.
- Every problem in our framework (i) is solvable in polynomial time if the dimension d is fixed (see Sect. 3.1), and (ii) allows a polynomial time 2-approximation algorithm (see Sect. 3.2).
- For several problems in the framework (among which assignment, spanning tree, s,t-cut, connecting path and Horn-SAT), we prove that the existence of an approximation algorithm with approximation ratio strictly better than 2 implies $P = NP$ (see Sect. 4.1). Note that these results pinpoint the strongest achievable approximation ratio for these problems (under $P \neq NP$).
- For one problem in our framework (2SAT) we prove that it is actually solvable in polynomial time (see Sect. 4.2). Thus, not all problems in the framework are NP-hard.
- For the balanced assignment problem with vector *sum* costs we prove that the existence of a polynomial time approximation algorithm with approximation ratio below $\frac{4}{3}$ implies $P = NP$; see Sect. 5.

2   The Framework

Throughout, we consider a family of optimization problems that are built around a finite ground set X and a system $\mathcal{F}$ of feasible subsets over X. (The system $\mathcal{F}$ is usually not listed explicitly, but given implicitly in terms of a combinatorial description or in terms of an oracle.) We will only consider problems in this framework, for which the following *feasibility oracle* can be performed in time polynomially bounded in the size of X: "Given a subset $Y \subseteq X$, does Y contain a feasible subset from $\mathcal{F}$? And if so return a feasible subset of Y from $\mathcal{F}$." Here are some concrete examples of problems that fit this framework:

q-**Uniform Set System.** For a given ground set X, a subset $Y \subseteq X$ is feasible if it contains at least q elements of X.

Linear Assignment. The ground set X are the elements of an $n \times n$ square matrix. A subset $Y \subseteq X$ is feasible if it contains n elements that cover each row and each column of the given matrix.

Spanning Tree. The ground set X consists of the edges of an undirected graph $G = (V, X)$. A subset $Y \subseteq X$ is feasible if the subgraph (V, Y) contains a spanning tree of G.

s, t-**Cut.** The ground set X consists of the edges of an undirected graph $G = (V, X)$ with $s, t \in V$. A subset $Y \subseteq X$ is feasible if it contains an s, t-cut; in other words, the subgraph $(V, E \setminus Y)$ contains no path connecting s and t.

Connecting Path. The ground set X consists of the edges of an undirected graph $G = (V, X)$ with $s, t \in V$. A subset $Y \subseteq X$ is feasible if the subgraph (V, Y) contains a path connecting s and t.

2SAT, Horn-SAT. The ground set X consists of all literals both positive and negated of an expression in conjunctive normal form, i.e. $X = \{x_1, \bar{x}_1, \ldots, x_n, \bar{x}_n\}$. A subset $Y \subseteq X$ is feasible if there exists a feasible assignment with the literals in Y. An assignment is feasible if each literal is set to either TRUE or FALSE (either x or $\bar{x}$ is in Y), such that all clauses in the expression are satisfied.

Here is an example of a problem that does NOT fall under this framework (unless $P = NP$):

Hamiltonicity. The ground set X consists of the edges of an undirected graph $G = (V, X)$. A subset $Y \subseteq X$ is feasible if the subgraph (V, Y) contains a Hamiltonian cycle.

We will study so-called *balanced vector-cost* versions of the problems in the framework. For this, we generalize the terminology introduced in Sect. 1 in the following way. Besides the ground set X and the system $\mathcal{F}$ of feasible subsets, we introduce a cost function $c : X \to \mathbb{R}^d$ that assigns to every element $x \in X$ of the ground set a corresponding d-dimensional real vector $\mathbf{c}(x)$; the d components of vector $\mathbf{c}(x)$ will be denoted $c_1(x), \ldots, c_d(x)$. For a subset $Y \subseteq X$, its *imbalance* in the k-th coordinate $(1 \le k \le d)$ is defined as:

$$\Delta_k(Y) \;=\; \max_{y \in Y} c_k(y) \;-\; \min_{y \in Y} c_k(y).$$

In other words, this imbalance measures the difference in cost between the largest and smallest value in the k-th coordinate. The *imbalance* of subset Y is finally defined as

$$\Delta_{\max}(Y) \;=\; \max_{1 \le k \le d} \Delta_k(Y).$$

The goal in a balanced vector-cost optimization problem is to find a feasible set Y that minimizes the imbalance $\Delta_{\max}(Y)$. In the sequel, the term "the balanced vector-cost problem" refers to an arbitrary problem in our framework.

3   Algorithms for Balanced Vector-Cost Problems

In this section we give two algorithms that are applicable to any problem in our general framework. The first algorithm solves the problem in polynomial time when the dimension d of the cost-vectors is fixed (Sect. 3.1). The second algorithm yields a 2-approximation in polynomial time (Sect. 3.2). We remind the reader that we only consider problems for which the feasibility oracle can be performed in polynomial time. Throughout this section we use $n := |X|$.

3.1   Fixed Dimension

We first explain the general idea behind Algorithm 1, which generalizes an algorithm presented in Martello et al. [10].

Suppose we would know the largest and smallest value in each coordinate of an optimal solution, without knowing the elements in the optimal solution. This allows us to construct a subset $Y \subseteq X$ consisting of elements whose cost-vectors, in every coordinate, lies within these values. Next, applying the feasibility oracle to Y gives us an optimum solution.

Of course, we are not given these values. However, one pair of elements of the ground set allows us to 'guess' the largest and smallest value in one coordinate. If the dimension d of the cost-vectors is fixed, it is sufficient to try all possible combinations of these 'guesses', as is argued in Theorem 1.

Algorithm 1

1: Sol $:= \infty$
2: **for** each $x_1, y_1 \in X$ **with** guess$_1 = [c_1(x_1), c_1(y_1)]$ **do**
3:     **for** each $x_2, y_2 \in X$ **with** guess$_2 = [c_2(x_2), c_2(y_2)]$ **do**

4:         $\vdots$

5:             **for** each $x_d, y_d \in X$ **with** guess$_d = [c_d(x_d), c_d(y_d)]$ **do**
6:                 Let $v := \max_k |$guess$_k|$
7:                 **for** x in X **do**
8:                     Remove x if $\exists k \in \{1, \ldots, d\}$ such that $c_k(x) \notin$ guess$_k$
9:                 Call the remaining set of elements Y
10:                **if** Y contains a feasible solution **then**
11:                    Sol $:= \min\{$Sol$, v\}$
12: Output Sol.

Theorem 1. *Algorithm 1 solves the balanced vector-cost problem in polynomial time, when the dimension d of the cost-vectors is fixed.*

Proof. Consider an optimal solution with value OPT, and let $\Delta_k(OPT)$ be its imbalance in the k-th coordinate $(1 \leq k \leq d)$. By trying out all combinations of guess$_k$ for each k, we will certainly find the imbalances $\Delta_k(OPT)$. Hence we will find a solution with value $v = \max_k \Delta_k(OPT) = OPT$, the optimal objective value.

In total there are $O(n^{2d})$ 'guesses'. For each of them we call the feasibility oracle to check whether there exists a feasible solution. These feasibility oracles run in polynomial time and hence the algorithm runs in polynomial time (for a fixed d).   □

3.2 An Approximation Algorithm

When the dimension d of the cost-vectors is part of the input the problem becomes more difficult. Simply trying all combinations of pairs for each coordinate would now result in an exponential time algorithm. Instead, we consider every pair of the ground set as a guess for *all* coordinates at the same time. Next, we will only consider elements from the ground set that, in every coordinate, do not differ 'too much' from this pair. Doing so gives us a 2-approximation, even when the dimension d of the cost-vectors is part of the input.

Algorithm 2

1: **for** each pair x_1, x_2 in ground set X **do**
2:      $\Delta_{x_1,x_2} = \max_k |c_k(x_1) - c_k(x_2)|$
3:      **for** x in X **do**
4:          **if** $\max_k |c_k(x_1) - c_k(x)| > \Delta_{x_1,x_2}$ **or** $\max_k |c_k(x_2) - c_k(x)| > \Delta_{x_1,x_2}$ **then**
5:              **Remove** x
6:      Call the remaining set of elements Y
7:      **if** Y contains a feasible solution **then**
8:          $\text{Sol}(x_1, x_2) := \Delta_{\max}(Y)$
9: $\text{Sol} := \min_{x_1,x_2} \text{Sol}(x_1, x_2)$

Theorem 2. *Algorithm 2 is a 2-approximation algorithm for the balanced vector-cost problem.*

Proof. Let OPT denote the imbalance of an optimal solution. By trying out all possible element pairs x_1 and x_2 from the ground set, we will certainly find the two elements in the optimal solution that determine the objective value; in other words, $\Delta_{\max}(\{x_1, x_2\}) = OPT$.

We remove all elements y from the ground set that satisfy $\Delta_{\max}(\{y, x_1\}) > \Delta$ or $\Delta_{\max}(\{y, x_2\}) > \Delta$; note that these removed elements can never show up in an optimal solution that contains x_1 and x_2 and that has imbalance Δ. Clearly, $\Delta_{\max}(Y)$ is determined by two of its elements, say y_1 and y_2. In other words, there exist $y_1, y_2 \in Y$ such that

$$\Delta_{\max}(Y) = \Delta_{\max}(\{y_1, y_2\}) \leq \Delta_{\max}(\{y_1, x_1\}) + \Delta_{\max}(\{y_2, x_1\}) \leq 2\Delta.$$

Clearly, this procedure runs in polynomial time: checking whether an element $x \in X$ needs to be removed can be done in $O(d)$ time, and we need to perform the feasibility oracle $O(n^2)$ times. □

Notice that this algorithm also applies to problems for which the feasibility oracle is not solvable in polynomial time. More precisely, let $f(n)$ denote the running time of the feasibility oracle. The running time of Algorithm 2 equals $O\left(n^3 \cdot (d + f(n))\right)$.

4   The Complexity of Balanced Vector-Cost Problems

Many balanced optimization problems with scalar costs are known to be solvable in polynomial time (see the discussion in Sect. 1.1): q-Uniform Set Systems, the Linear Assignment problem, the Spanning Tree problem, the s,t-Cut problem, the Connecting Path problem, Horn-SAT and 2SAT. In this section we discuss the complexity of each of these problems when vector costs are given. We show that each of these problems, except 2SAT, is NP-hard, and that the existence of a polynomial-time $(2 - \epsilon)$-approximation algorithm for each of the mentioned problems, except 2SAT, implies $P = NP$ (Sect. 4.1). We also show that 2SAT is in fact polynomial solvable, which proves that not all problems in the framework are NP-hard (Sect. 4.2).

There are three problems, well-known to be NP-complete, that we use in our reductions.

Problem: INDEPENDENT SET (IS)

Instance: A graph $G = (V, E)$ with vertex set $V = \{v_1, \ldots, v_n\}$ and edge set $E = \{e_1, \ldots, e_m\}$; an integer z.

Question: Does there exist a subset $I \subseteq V$ with $|I| \geq z$, such that the vertices in I do not span any edges in G?

Problem: 3-COLORING

Instance: A graph $G = (V, E)$ with vertex set $V = \{v_1, \ldots, v_n\}$ and edge set $E = \{e_1, \ldots, e_m\}$.

Question: Does there exist a 3-coloring $f : V \rightarrow \{1, 2, 3\}$, such that all edges $(u, v) \in E$ satisfy $f(u) \neq f(v)$?

Problem: 3SAT

Instance: Set U of variables, collection C of clauses over U such that each clause in C contains 3 literals.

Question: Does there exist a satisfying truth assignment for C?

4.1 NP-Hardness Results

Let us first consider the balanced q-Uniform Set System with vector costs. Given a ground set X and an integer q, the balanced q-Uniform Set System with vector costs asks for q elements from set X with minimal imbalance.

Theorem 3. *The balanced q-Uniform Set System with vector costs is NP-hard.*

Proof. Given an instance of INDEPENDENT SET represented by a given graph $G = (V, E)$ and an integer z, we construct an instance of the balanced q-Uniform Set System with vector costs as follows. The ground set X coincides with the vertex set V of the graph G. A subset $Y \subseteq X$ is feasible if and only if it contains at least $q := z$ elements. For the definition of the vector costs of X, we turn G into a directed graph by first choosing some ordering of the vertices in V, and next orienting every edge from the incident vertex with smaller index (source) to the incident vertex with larger index (target). The dimension of the vectors is $d := |E| = m$, and every coordinate k corresponds to a unique edge e_k in E, $1 \le k \le m$. Let us now define cost-vector $\mathbf{c}(v_j) = (c_1(v_j), c_2(v_j), \dots, c_d(v_j))$ corresponding to each vertex $v_j \in V$. For each $v_j \in V$ and $k \in \{1, \dots, m\}$:

$$c_k(v_j) := \begin{cases} 1 & \text{if vertex } v_j \text{ is the source of the oriented edge } e_k; \\ -1 & \text{if vertex } v_j \text{ is the target of the oriented edge } e_k; \\ 0 & \text{otherwise.} \end{cases}$$

We claim that there exists a feasible subset $Y \subseteq X$ with $|Y| \ge z$ and $\Delta_{\max}(Y) \le 1$ if and only if the considered instance of INDEPENDENT SET has answer YES.

Assume that there exists a feasible subset $Y \subseteq X$ with $|Y| \ge z$ and $\Delta_{\max}(Y) \le 1$. Suppose for the sake of contradiction that the vertex set corresponding to Y would span some edge $e_k \in E$. Then, in the k-th coordinate, the cost-vector of the source vertex of e_k is -1, and the cost-vector of the target vertex of e_k is $+1$. Hence $\Delta_{\max}(Y) \ge 2$. This contradiction shows that Y is a z-element independent set in G.

Next assume that the INDEPENDENT SET instance has answer YES and let I be the corresponding certificate. Thus $|I| \ge z$ and in none of the coordinates, the vectors $\mathbf{c}(y)$ with $y \in I$ take the value $+1$ and -1. This yields the desired $\Delta_{\max}(I) \le 1$. □

Theorem 4. *The balanced q-Uniform Set System with vector costs does not allow a polynomial time approximation algorithm with worst case guarantee strictly better than 2 (unless $P = NP$).*

Proof. This is implied by the proof of Theorem 3. Indeed, a polynomial time approximation algorithm with a worst case guarantee strictly better than 2, would allow us to distinguish the instances with imbalance at most 1 from the instances with imbalance at least 2. □

Consider the following 5 specific balanced optimization problems with vector costs:

Linear Assignment: given a square matrix M, where each entry is a d-dimensional vector, the goal is to find an assignment in M with minimal imbalance.

Spanning Tree: given a graph with a cost-vector for each edge, the goal is to find a spanning tree with minimal imbalance.

s, t-Cut: given a graph with a cost-vector for each edge, and two nodes s and t, the goal is to find a cut in this graph separating s and t with minimal imbalance.

Connecting Path: given a graph with a cost-vector for each edge, and two nodes s and t, the goal is to find a path connecting s and t with minimal imbalance.

Horn-SAT: given a set of literals X with a cost-vector for each literal and a set of clauses C', each clause with at most 1 positive literal, the goal is to find a satisfying truth assignment with minimal imbalance.

We denote this set of problems by set Q. The theorem below is proven for each problem separately and can be found in the corresponding research report.

Theorem 5. *Each of these problems in Q is NP-hard and moreover, no polynomial time approximation algorithm exists with worst case guarantee strictly better than 2 (unless $P = NP$).*

4.2 2SAT

Given a set of literals $X = \{x_1, \bar{x}_1, \ldots, x_n, \bar{x}_n\}$ with a cost-vector for each literal and a set of clauses $C = \{c_1, \ldots, c_m\}$ each with at most 2 literals, the balanced 2SAT problem with vector costs asks to find a satisfying truth assignment which minimizes the imbalance. We show that this problem, unlike the previous problems, is easy.

Theorem 6. *The balanced 2SAT problem with vector costs is polynomial solvable.*

Proof. First, we prove that we can decide in polynomial time whether a solution with imbalance Δ exists. Consider the cost-vectors of each pair of elements u, $v \in X$. If there is a coordinate in which these two vectors differ more than Δ, then these two elements cannot occur together in a solution with imbalance Δ. Hence we add the clause $(\bar{u} \vee \bar{v})$ to the set of clauses of each such pair; where the negation of a negated literal results in a positive literal, i.e. $\bar{\bar{x}} = x$. Notice that the 2SAT instance remains a 2SAT instance and that each feasible solution to this new instance is a feasible solution to the original problem with balance at most Δ.

We know that in any feasible solution its imbalance Δ is defined by two elements of the ground set X. That gives us at most $O(n^2)$ distinct possible values for Δ (one for each pair of elements). The lowest value of Δ for which there exists a truth assignment is the value of the optimal solution. Next, a binary search on Δ allows us to do a polynomial number of iterations.   □

5   A Special Case of the Balanced Assignment Problem with Vector-Costs: Sum Costs

Kamura and Nakamori [6] consider a highly structured special case of the balanced assignment problem with vector costs: the cost-vector for every matrix entry $M[a, b]$ is the sum of two d-dimensional cost-vectors $\mathbf{c}(a)$ and $\mathbf{c}(b)$. We call this setting the balanced assignment problem with vector *sum* costs. The resulting problem remains NP-hard, as witnessed by the following result.

The two theorems below are proven via a reduction from Independent Set, which is similar to the reduction of q-Uniform Set Systems. Both proofs can be found in the corresponding research report.

Theorem 7. *The balanced assignment problem with vector sum costs is NP-hard.*

Unfortunately this construction does not close the gap between the factor of 2 achieved by Algorithm 2, and what might be achieved by any polynomial time algorithm. We can only state:

Theorem 8. *The balanced assignment problem with sum vector costs does not allow a polynomial time approximation algorithm with worst case guarantee strictly better than $\frac{4}{3}$ (unless $P = NP$).*

Remark: Given the application described in Kamura and Nakamori [6], one could be interested in the balanced 3-dimensional assignment problem with vector costs. In this problem, we are given three sets of vectors, say a set A, B and C. Then, the ground set X consists of triples, each consisting of a vector from A, a vector from B, and a vector from C, and the costs of an element from X is nothing else but the sum of the three vectors. Although this problem does not fall in our framework (the feasibility question is NP-hard), one might wonder about the approximability of this balanced 3-dimensional assignment problem with vector costs. We point out, however, that no constant-factor approximation algorithm can exist (unless $P = NP$), even when $d = 1$.

Theorem 9. *The balanced 3-dimensional assignment problem with sum vector costs does not allow a polynomial time constant-factor approximation algorithm (unless $P = NP$), even when $d = 1$.*

Proof. There is a straightforward reduction from Numerical 3-Dimensional Matching. Recall that in Numerical 3-Dimensional Matching we are given disjoint sets W, Y and Z, each containing n elements and a cost $c(a)$ for each $a \in W \cup Y \cup Z$ and a bound b. The goal is to select n pairwise disjoint triples from $W \times Y \times Z$, referred to as M, such that for each of the selected triples, (w, y, z), it holds that $c(w) + c(y) + c(z) = b$.

By having an element in A (B, C), for each element in W (Y, Z) an equivalent instance of the balanced 3-dimensional assignment problem with sum vector costs arises. Notice that the imbalance in this instance is 0 if and only if there exists a set of triples M such that for each triple (w, y, z) in M it holds that $c(w) + c(y) + c(z) = b$. Distinguishing in polynomial time whether the imbalance is 0 or not would imply $P = NP$.   □

6   Conclusion

We introduce the notion of balanced vector-cost optimization problems, and propose a framework that generalizes the one introduced by Martello et al. [10]. We provide a polynomial time algorithm when the dimension d is fixed, and we describe an algorithm that is a 2-approximation for each problem in our framework. Further, we give results for five problems in the framework: each of them is NP-hard, and the existence of a polynomial time $(2 - \epsilon)$-approximation algorithm implies P = NP.

Acknowledgements. This research has been supported by the Netherlands Organisation for Scientific Research (NWO) under Grant 639.033.403, by BSIK Grant 03018 (BRICKS: Basic Research in Informatics for Creating the Knowledge Society), and by the Interuniversity Attraction Poles Programme initiated by the Belgian Science Policy Office.

References

1. Ahuja, R.: The balanced linear programming problem. Eur. J. Oper. Res. **101**(1), 29–38 (1997)
2. Camerini, P., Maffioli, F., Martello, S., Toth, P.: Most and least uniform spanning trees. Discrete Appl. Math. **15**(23), 181–197 (1986)
3. Cappanera, P., Scutellà, M.G.: Balanced paths in acyclic networks: tractable cases and related approaches. Networks **45**(2), 104–111 (2005)
4. Dokka, T., Crama, Y., Spieksma, F.C.R.: Multi-dimensional vector assignment problems. Discrete Optim. **14**, 111–125 (2014)
5. Galil, Z., Schieber, B.: On finding most uniform spanning trees. Discrete Appl. Math. **20**(2), 173–175 (1988)
6. Kamura, Y., Nakamori, M.: Modified balanced assignment problem in vector case: system construction problem. In: 2014 International Conference on Computational Science and Computational Intelligence (CSCI), vol. 2, pp. 52–56. IEEE (2014)
7. Katoh, N., Iwano, K.: Efficient algorithms for minimum range cut problems. Networks **24**(7), 395–407 (1994)
8. Kinable, J., Smeulders, B., Delcour, E., Spieksma, F.C.R.: Exact algorithms for the Equitable Traveling Salesman Problem. Research report, KU Leuven (2016)
9. Larusic, J., Punnen, A.: The balanced traveling salesman problem. Comput. Oper. Res. **38**(5), 868–875 (2011)
10. Martello, S., Pulleyblank, W., Toth, P., De Werra, D.: Balanced optimization problems. Oper. Res. Lett. **3**(5), 275–278 (1984)
11. Turner, L.: Variants of shortest path problems. Algorithmic Oper. Res. **6**(2), 91–104 (2012)

Vertex Sparsification in Trees

Gramoz Goranci[1]([✉]) and Harald Räcke[2]

[1] Faculty of Computer Science, University of Vienna, Vienna, Austria
gramoz.goranci@univie.ac.at
[2] Institut für Informatik, Technische Universität München, Garching, Germany
raecke@in.tum.de

Abstract. Given an unweighted tree $T = (V, E)$ with terminals $K \subset V$, we show how to obtain a 2-quality vertex flow and cut sparsifier H with $V_H = K$. We prove that our result is essentially tight by providing a $2 - o(1)$ lower-bound on the quality of any cut sparsifier for stars.

In addition we give improved results for quasi-bipartite graphs. First, we show how to obtain a 2-quality flow sparsifier with $V_H = K$ for such graphs. We then consider the other extreme and construct exact sparsifiers of size $O(2^k)$, when the input graph is unweighted.

Keywords: Graph sparsification · Vertex flow sparsifiers · Trees

1 Introduction

Graph sparsification is a technique to deal with large input graphs by "compressing" them into smaller graphs while preserving important characteristics, like cut values, graph spectrum etc. Its algorithmic value is apparent, since these smaller representations can be computed in a preprocessing step of an algorithm, thereby greatly improving performance.

Cut sparsifiers [4] and spectral sparsifiers [19] aim at reducing the number of edges of the graph while approximately preserving cut values and graph spectrum, respectively. These techniques are used in a variety of fast approximation algorithms, and are instrumental in the development of nearly linear time algorithms.

In vertex sparsification [6,9,10,12,14,16,18], apart from reducing the number of edges, the goal is also to reduce the number of vertices of a graph. In such setting, one is given a large graph $G = (V, E, c)$, together with a relatively small subset of terminals $K \subseteq V$. The goal is to shrink the graph while preserving properties involving the terminals. For example, in *Cut Sparsification* one wants to construct a graph $H = (V_H, E_H, c_H)$ (with $K \subseteq V_H$) such that H preserves mincuts between terminals up to some approximation factor q (the *quality*).

Hagerup et al. [9] introduced this concept under the term *Mimicking Networks*, and focused on constructing a (small) graph H that maintains mincuts exactly. They showed that one can obtain H with $O(2^{2^k})$ vertices, where $k = |K|$. Krauthgamer and Rika [13] and Khan and Raghavendra [11] independently proved that $2^{\Omega(k)}$ vertices are required for some graphs if we want to preserve mincuts exactly.

© Springer International Publishing AG 2017
K. Jansen and M. Mastrolilli (Eds.): WAOA 2016, LNCS 10138, pp. 103–115, 2017.
DOI: 10.1007/978-3-319-51741-4_9

Moitra [16] analyzed the setting where the graph H is as small as possible, namely $V_H = K$. Under this condition, he obtained a quality $O(\log k / \log \log k)$ cut sparsifier. A lower bound of $\Omega(\sqrt{\log k} / \log \log k)$ was presented by Makarychev and Makarychev [15]. A strictly stronger notion than a cut sparsifier, is a *flow sparsifier* that aims at (approximately) preserving all multicommodity flows between terminals. The upper bound of [16] also holds for this version, but the lower bound is slightly stronger: $\Omega(\sqrt{\log k} / \log \log k)$.

Due to the lower bounds on the quality of sparsifiers with $V_H = K$, the recent focus has been on obtaining better guarantees with slightly larger sparsifiers. Chuzhoy [7] obtained a constant quality flow sparsifier of size $C^{O(\log \log C)}$, where C is the total weight of the edges incident to terminal nodes. Andoni et al. [3] obtained quality of $(1 + \varepsilon)$ and size $O(\text{poly}(k/\varepsilon))$ for *quasi-bipartite* graphs, i.e., graphs where the terminals form an independent set. This is interesting since these graphs serve as a lower bound example for Mimicking Networks, i.e., in order to obtain an exact sparsifier one needs size at least $2^{\Omega(k)}$.

In this paper we study flow and cut sparsifiers for trees. Since, for tree networks it is immediate to obtain a sparsifier of size $O(k)$ and quality 1, we consider the problem of designing flow and cut sparsifiers with $V_H = K$ as in the original definition of Moitra. In Sect. 2 we show how to design such a flow sparsifier for unweighted trees with quality 2. In Sect. 3 we prove that this result is essentially tight by establishing a lower bound. Concretely, we prove that even for unweighted stars it is not possible to obtain cut sparsifiers with quality $2 - o(1)$.

As a further applicaton of our techniques, we apply them to quasi-bipartite graphs. We first obtain a 2-quality flow sparsifier with $V_H = K$ for such graphs. In addition we explore the other extreme and construct exact sparsifiers of size $O(2^k)$, if the input graph is unweighted. This shows that even though quasi-bipartite graphs serve as lower bound instances for Mimicking Networks they are not able to close the currently large gap between the upper bound of $O(2^{2^k})$ and the lower bound of $2^{\Omega(k)}$ on the size of Mimicking Networks.

Finally we obtain hardness results for the problem of deciding whether a graph H is a sparsifier for a given unweighted tree T. We prove that this problem is co-$\mathcal{NP}$-hard for cut sparsifiers, based on Chekuri et al. [5]. For flow sparsifiers we show that for a single-source version, where the sparsifier has to preserve flows in which all demands share a common source, the problem is co-$\mathcal{NP}$-hard. Due to space limitations the hardness results have been deferred to the full version.

1.1   Preliminaries

Let $G = (V, E, c)$ be an undirected graph with terminal set $K \subset V$ of cardinality k, where $c : E \to \mathbb{R}^+$ assigns a non-negative capacity to each edge. We present two different ways to sparsify the number of vertices in G.

Let $U \subset V$ and $S \subset K$. We say that a cut $(U, V \setminus U)$ is S-separating if it separates the terminal subset S from its complement $K \setminus S$, i.e., $U \cap K$ is either S or $K \setminus S$. The cutset $\delta(U)$ of a cut $(U, V \setminus U)$ represents the edges that have one endpoint in U and the other one in $V \setminus U$. The cost $\text{cap}_G(\delta(U))$ of a cut $(U, V \setminus U)$ is the sum over all capacities of the edges belonging to the cutset.

We let $\text{mincut}_G(S, K\setminus S)$ denote the S-separating cut of the minimum cost in G. A graph $H = (V_H, E_H, c_H)$, $K \subset V_H$ is a *vertex cut sparsifier* of G with *quality* $q \geq 1$ if: $\forall S \subset K$, $\text{mincut}_G(S, K\setminus S) \leq \text{mincut}_H(S, K\setminus S) \leq q\cdot\text{mincut}_G(S, K\setminus S)$.

We say that a *(multi-commodity)* flow f is a *routing* of the demand function d, if for every terminal pair (x, x') it sends $d(x, x')$ units of flow from x to x'. The *congestion* of an edge $e \in E$ incurred by the flow f is defined as the ratio of the total flow sent along the edge to the capacity of that edge, i.e., $f(e)/c(e)$. The *congestion of the flow f* for routing demand d is the maximum congestion over all edges in G. We let $\text{cong}_G(d)$ denote the minimum congestion over all flows. A graph $H = (V_H, E_H, c_H)$, $K \subset V_H$ is a *vertex flow sparsifier* of G with *quality* $q \geq 1$ if for every demand function d, $\text{cong}_H(d) \leq \text{cong}_G(d) \leq q \cdot \text{cong}_H(d)$.

We use the following tools about sparsifiers throughout the paper.

Lemma 1 [14]. *If $H = (V_H, E_H, c_H)$, $V_H = K$ is a vertex flow sparsifier of G, then the quality of H is $q = \text{cong}_G(d_H)$, where $d_H(x, x') := c_H(x, x')$ for all terminal pairs (x, x').*

Let G_1 and G_2 be graphs on *disjoint* set of vertices with terminals $K_1 = \{s_1, \ldots, s_k\}$ and $K_2 = \{t_1, \ldots, t_m\}$, respectively. In addition, let $\phi(s_i) = t_i$, for all $i = 1, \ldots, \ell$, be a one-to-one correspondence between some subset of K_1 and K_2. The *ϕ-merge* (or *2-sum*) of G_1 and G_2 is the graph G with terminal set $K = K_1 \cup \{t_{\ell+1}, \ldots, t_m\}$ formed by identifying the terminals s_i and t_i for all $i = 1, \ldots, \ell$. This operation is denoted by $G := G_1 \oplus_\phi G_2$.

Lemma 2 ([3], **Merging**). *Let $G = G_1 \oplus_\phi G_2$. Suppose G_1' and G_2' are flow sparsifiers of quality q_1 and q_2 for G_1 and G_2, respectively. Then $G' = G_1' \oplus_\phi G_2'$ is a flow sparsifier of quality $\max\{q_1, q_2\}$ for G.*

Lemma 3 (**Convex Combination of Sparsifiers**). *Let $H_i = (V^*, E_i, c_i)$, $i = 1, \ldots, m$ with $K \subset V^*$ be vertex flow sparsifiers of G. In addition, let $\alpha_1, \alpha_2, \ldots, \alpha_m$ be convex multipliers corresponding to H_i's such that $\sum_i \alpha_i = 1$. Then the graph $H' = \sum_i \alpha_i \cdot H_i$ is a vertex flow sparsifier for G.*

2   Improved Vertex Flow Sparsifiers for Trees

In this section we show that given an unweighted tree $T = (V, E)$, $K \subset V$, we can construct a flow sparsifier H only on the terminals, i.e., $V(H) = K$, with quality at most 4. We then further improve the quality to 2. The graph H has the nice property of being a convex combination of trees.

We obtain the quality of 4 by combining the notion of probabilistic mappings due to Andersen and Feige [2] and a duality argument due to Räcke [17]. Our result then immediately follows using as a black-box an implicit result of Gupta [8]. We note that a direct application of the Transfer Theorem due to Andersen and Feige [2] does not apply, since their interchangeability argument relies on arbitrary capacities and lengths.

Let $w : E \to \mathbb{R}_{\geq 0}$ be a function which assigns non-negative values to edges which we refer to as *lengths*. Given a tree $T = (V, E, w)$ we use $d_w : V \times V \to \mathbb{R}_{\geq 0}$

to denote the shortest path distance induced by the edge length w. A *0-extension* of a tree $T = (V, E)$, $K \subset V$ is a retraction $f : V \to K$ with $f(x) = x$, for all $x \in K$, along with another graph $H = (K, E_H)$ such that $E_H = \{(f(u), f(v)) : (u, v) \in E\}$. The graph H is referred to as a *connected 0-extension* if in addition we require that $f^{-1}(x)$ induces a connected component in T.

Given a graph $G = (V, E)$, we let $\mathcal{P}$ be a collection of multisets of E, which will be usually referred to as paths. A mapping $M : E \to \mathcal{P}$ maps every edge e to a path $P \in \mathcal{P}$. This mapping can be alternatively represented as a non-negative square matrix M of dimension $|E| \times |E|$, where $M(e', e)$ is the number of times edge e lies on the path $M(e')$. Let $\mathcal{M}$ denote the collection of mappings M. If we associate to each mapping $M \in \mathcal{M}$ a convex multiplier λ_M, the resulting mapping is referred to as a *probabilistic mapping*.

Connected 0-Extension Embedding on Trees. Suppose we are given a tree $T = (V, E)$, $K \subset V$ and a connected 0-extension (H, f), where $H = (K, E_H)$ and f is a retraction. Given an edge $(u, v) \in E$ from T, we can use the retraction f to find the edge $(f(u), f(v))$ in H (if u and v belong to different components). Since this edge is not an edge of the original tree T, we need a way to map it back to T in order to be consistent with our definition of mappings. The natural thing to do is to take the unique shortest path between $f(u)$ and $f(v)$ in T. Denote by $S_{u,v}^T$ all the edges in the shortest path between u and v in T. Then, we let $M_{H,f}((u, v)) = S_{f(u),f(v)}^T$ be the mapping $M_{H,f} : E \to \mathcal{P}$ induced by (H, f).

Let $\mathcal{H}$ be the family of all connected 0-extensions for T, which are also trees. We then define the collection of mappings $\mathcal{M}$ for T by $\{M_{H,f} : H \in \mathcal{H}\}$.

Capacity Mappings. Given a tree $T = (V, E, c)$, $c : E \to \mathbb{R}^+$ and a connected 0-extension (H, f), the *load* of an edge $e \in E$ under (H, f) is $\text{load}_f(e) = \sum_{e'} M_{H,f}(e', e) \cdot c(e')$. The *expected load* of an edge $e \in E$ under a probabilistic mapping is $\sum_i \lambda_i \text{load}_{f_i}(e)$.

Distance Mappings. Given a tree $T = (V, E, w)$, $w : E \to \mathbb{R}^+$ and a connected 0-extension (H, f), the *mapped length* of an edge $e' = (u', v') \in E$ under (H, f) is $d_w(f(u'), f(v')) = \sum_e M_{H,f}(e', e) \cdot w(e)$. The *expected mapped length* of an edge $e' = (u', v') \in E$ under a probabilistic mapping is $\sum_i \lambda_i d_w(f_i(u'), f_i(v'))$.

With the above definitions in mind, for some given tree $T = (V, E, c)$, we can find a flow sparsifiers that is a convex combination of connected 0-extensions using the following linear program, and its dual.

min α

s.t. $\forall e \sum_i \lambda_i \cdot \text{load}_{f_i}(e) \leq \alpha \cdot c(e)$

$\sum_i \lambda_i \geq 1$

$\forall i \qquad \lambda_i \geq 0.$

min β

s.t. $\forall i \sum_e w(e) \cdot \text{load}_{f_i}(e) \geq \beta$ $(*)$

$\sum_e w(e) \cdot c(e) \leq 1$

$\forall e \qquad w(e) \geq 0.$

Next, we re-write the dual constraints of type $(*)$ as follows:

$$\sum_e w(e) \text{load}_{f_i}(e) = \sum_e w(e) \sum_{e'} M_{H,f_i}(e', e) \cdot c(e')$$
$$= \sum_{e'} c(e') \left(\sum_e M_{H,f_i}(e', e) \cdot w(e) \right) = \sum_{e'=(u',v')} c(e') \cdot d_w(f_i(u'), f_i(v')).$$

Using this re-formulation and a few observations, the dual is equivalent to:

$$\max_{w \geq 0} \min_{i} \sum_{e=(u,v)} c(e) \cdot d_w(f_i(u), f_i(v)) \Big/ \sum_{e} w(e) \cdot c(e). \qquad (1)$$

For the unweighted case $c(e) = 1$, we can make use of the following lemma:

Lemma 4 [8, Lemma 5.1]. *Given a tree $T = (V, E, w)$, $K \subset V$, we can find a connected 0-extension f such that $\sum_{e=(u,v)} d_w(f(u), f(v)) \leq 4 \cdot \sum_e w_e$.*

The above lemma tells us that optimal value of (1) is bounded by 4. This implies that the optimal value of the dual is bounded by 4, and by strong duality, the optimal value of the primal is also bounded by 4. The latter implies that T admits a 4-quality vertex sparsifier of size k.

2.1   Obtaining Quality 2

Next we show how to bring down the quality of flow sparsifiers on trees to 2. We give a direct algorithm that constructs a flow sparsifiers and unlike in the previous subsection, it does not rely on the interchangeability between distances and capacities. We first consider trees where terminals are the only leaf nodes, i.e., $L(T) = K$. Later we show how to extend the result to arbitrary trees.

   To convey some intuition, we start by presenting the deterministic version of our algorithm. We maintain at any point of time a partial mapping f–setting $f(v) = \perp$, when $f(v)$ is still undefined, but producing a valid connected 0-extension when the algorithm terminates. Note that $f(x) = x$, for all $x \in K$. Without loss of generality, we may assume that the tree is rooted at some non-terminal vertex and the child-parent relationships are defined. The algorithm works as follows: it repeatedly picks a non-terminal v farthest from the root and maps it to one of its children c, i.e., $f(v) = f(c)^1$ (we refer to such procedure as Algorithm 1. This process results in a flow sparsifier that is a connected 0-extension.

   Unfortunately, the quality of the sparsifier produced by the above algorithm can be very poor. To see this, consider an unweighted star graph $S_{1,k}$, where leaves are the terminal vertices and the center is the non-terminal vertex v. Any connected 0-extension of $S_{1,k}$ is a new star graph $S_{1,k-1}$ lying on the terminals, where the center is the terminal x with $f(v) = x$. Now, consider a demand function d that sends a unit flow among all edges in $S_{1,k-1}$. Clearly, d can be feasibly routed in $S_{1,k-1}$. But routing d in $S_{1,k}$ gives a load of at least $k-1$ along the edge (x, v), and thus the quality of $S_{1,k-1}$ is at least $k-1$ (Lemma 1).

   One way to improve upon the quality is to map the non-terminal v uniformly at random to one of the terminals. We can equivalently view this as taking convex combination over all possible connected 0-extensions of $S_{1,k}$. By Lemma 3 we know that such a convex combination gives us another flow sparsifier for $S_{1,k}$, and it can be checked that the quality of such a sparsifier improves to 2. Surprisingly, we show that applying this trivial random-mapping of non-terminals in

---

[1] Alternatively, one can view this step as contracting an arbitrary child-edge of v.

trees with terminals as leaves leads to a flow sparsifier H which is a random connected 0-extension and achieves similar guarantees. We refer to such procedure
as Algorithm 2.

To compute the quality of H as a flow sparsifier for T, we need to bound
the congestion of every edge of T incurred by the embedding of H into T. This
embedding routes the capacity of every terminal edge (x, x') in H along the
(unique) shortest paths between leaves x and x' in T. First, we crucially observe
that without loss of generality, it suffices to bound the load of the edges incident
to the terminals, i.e., edges incident to leaf vertices. To see this, let (u, v) be
an edge among non-terminals in T, with v being the parent of u. Now, when
embedding H into T, we know that the demands among all terminal pairs that
lie in the subtree $T(u)$ rooted at u *cannot* incur any load on the edge (u, v), as
these terminal shortest paths do not use this edge. Thus, we can safely replace
the subtree $T(u)$ with some dummy terminal and perform the analysis as before.

First, we study edge loads under deterministic connected 0-extensions. Let
$e = (x, v)$ be the edge incident to $x \in K$, m_x denote the level of x in T and
$\{x, v_{m_x-1}, \ldots, v_0\}$ be the set of vertices belonging to the shortest path between
x and the root $r = v_0$ in T. Given a connected 0-extension f_i output by Algorithm 1, we say that x is *expanded* up to the ℓ-th level if $f_i(v_j) = x$, for all
$j \in \{m_x, \ldots, \ell\}$. This leads to the following lemma.

Lemma 5. *Let $e = (x, v)$ be the edge incident to $x \in K$, (H_i, f_i) be a connected
0-extension and recall that empty sum is defined as 0. If x is expanded up to the
ℓ-th level, then the load of e under (H_i, f_i) is* $\mathrm{load}_{f_i}(e) \leq 1 + \sum_{j=\ell}^{m_x-1}(c_j - 1)$, $\ell \in
\{m_x, \ldots, 0\}$, *where c_j denotes the number of children of non-terminal v_j in T.*

Let $I_\ell^x = \{(H_i, f_i)\}$ be the set of connected 0-extensions output by Algorithm 1
where x is expanded up to the ℓ-th level. We observe that the edge e has the same
load regardless of which element of I_ℓ^x we choose. Thus, for any $(H_i, f_i) \in I_\ell^x$,
we can write $\mathrm{load}_\ell(e) = \mathrm{load}_{f_i}(e)$.

Now, we study the expected edge loads under the random connected 0-
extension output by Algorithm 2. Let N be the number of all different connected
0-extensions that can be output by Algorithm 1. If by Z_ℓ^x we denote the event
that x is expanded up to the ℓ-th level, then it follows that the expected load
$\mathbb{E}[\mathrm{load}_f(e)]$ of $e = (x, v)$ under (H, f) is

$$\sum_{i=1}^{N} \mathrm{load}_{f_i}(e)/N = \sum_{\ell=0}^{m_x} \frac{\# \text{ of } f_i\text{'s s.t. } Z_\ell^x}{N} \cdot \mathrm{load}_\ell(e) = \sum_{\ell=0}^{m_x} \mathbb{P}[Z_\ell^x] \cdot \mathrm{load}_\ell(e). \quad (2)$$

Since in Algorithm 2 all non-terminals are mapped independently of each other,
we obtain $\mathbb{P}[Z_\ell^x] = (1 - 1/c_{\ell-1}) \prod_{j=\ell}^{m_x-1} 1/c_j, \ell \in \{m_x, \ldots, 1\}$ (recall that the
empty product is defined as 1). Further, observe that $\mathbb{P}[Z_0^x] = 1/\prod_{j=0}^{m_x-1} c_j$.
Plugging the probabilities and Lemma 5 in (2), we get that $\mathbb{E}[\mathrm{load}_f(e)]$ is

$$\frac{1}{\prod_{j=0}^{m_x-1} c_j}\left(1 + \sum_{j=0}^{m_x-1}(c_j - 1)\right) + \sum_{\ell=1}^{m_x}(1 - 1/c_{\ell-1}) \prod_{j=\ell}^{m_x-1} \frac{1}{c_j}\left(1 + \sum_{j=\ell}^{m_x-1}(c_j - 1)\right).$$

Next, we rewrite the above as A/B, where $B = \prod_{j=0}^{m_x-1} c_j$ and A is given by

$$1 + \sum_{j=0}^{m_x-1}(c_j - 1) + \sum_{\ell=1}^{m_x-1}(c_{\ell-1}-1)\prod_{j=0}^{\ell-2}c_j\left(1 + \sum_{j=\ell}^{m_x-1}(c_j - 1)\right) + (c_{m_x-1}-1)\prod_{j=0}^{m_x-2}c_j.$$

The following lemma simplifies the middle expression of A.

Lemma 6. *For any positive integers $\{c_0,\dots,c_{m_x-1}\}$ and $m_x \geq 3$,*

$$\sum_{\ell=1}^{m_x-1}(c_{\ell-1}-1)\prod_{j=0}^{\ell-2}c_j\left(1 + \sum_{j=\ell}^{m_x-1}(c_j - 1)\right) = (c_{m_x-1}+1)\prod_{\ell=0}^{m_x-2}c_\ell - \sum_{\ell=0}^{m_x-1}(c_\ell - 1) - 2.$$

Proof. Let $P(m_x - 1)$ be the left-hand side expression in the statement of the lemma. We proceed by induction on m_x. For the base case $m_x = 3$, it is easy to argue that the claim is valid. If we assume that the lemma holds true for $m_x - 1$, then we get that:

$$
\begin{aligned}
P(m_x) &= \sum_{\ell=1}^{m_x-1}(c_{\ell-1}-1)\prod_{j=0}^{\ell-2}c_j\left(1 + \sum_{j=\ell}^{m_x-1}(c_j-1) + (c_{m_x}-1)\right) \\
&\quad + (c_{m_x-1}-1)\prod_{j=0}^{m_x-2}c_j\left((c_{m_x}-1)+1\right) \\
&= \sum_{\ell=1}^{m_x-1}(c_{\ell-1}-1)\prod_{j=0}^{\ell-2}c_j\left(1 + \sum_{j=\ell}^{m_x-1}(c_j-1)\right) \\
&\quad + (c_{m_x}-1)\sum_{\ell=1}^{m_x}(c_{\ell-1}-1)\prod_{j=0}^{\ell-2}c_j + (c_{m_x-1}-1)\prod_{j=0}^{m_x-2}c_j \ .
\end{aligned}
$$ (3)

Note that the following expression is a simple telescoping series:

$$\sum_{\ell=1}^{m_x}(c_{\ell-1}-1)\prod_{j=0}^{\ell-2}c_j = \prod_{\ell=0}^{m_x-1}c_\ell - 1.$$ (4)

Plugging this into Eq. (3) and using induction hypothesis gives:

$$
\begin{aligned}
P(m) &= (c_{m_x-1}+1)\prod_{\ell=0}^{m_x-2}c_\ell - \sum_{\ell=0}^{m_x-1}(c_\ell-1) - 2 \ \ + (c_{m_x}-1)\left(\prod_{\ell=0}^{m_x-1}c_\ell - 1\right) \\
&\quad + (c_{m_x-1}-1)\prod_{j=0}^{m_x-2}c_j = (c_{m_x}+1)\prod_{\ell=0}^{m_x-1}c_\ell - \sum_{\ell=0}^{m_x}(c_\ell-1) - 2.
\end{aligned}
$$

This completes the induction step, and hence the proof of the lemma.   □

Now, plugging the above lemma in A we get that $A = 2B - 1$. Thus, $\mathbb{E}[\mathrm{load}_f(e)] = (2B - 1)/B \leq 2$. Since we consider only unweighted trees, it follows that the expected congestion of every edge is also bounded by 2. Taking the maximum over all edge congestions yields the following:

Lemma 7. *Given a tree* $T = (V, E)$, $K \subset V$, $L(T) = K$, *there is a 2-quality flow sparsifier* H, *which is a convex combination over connected 0-extensions.*

Derandomization. Next we show that Algorithm 2 can be easily derandomized. We obtain a deterministic algorithm that runs $O(n + k^2\alpha(2k))$ time and gives the same guarantees as in Lemma 7, where $\alpha(\cdot)$ is the inverse Ackermann function.

We first give an $O(n)$ time preprocessing step. For a tree $T = (V, E)$, $K \subset V$, $L(T) = K$, we repeatedly contract edges incident to non-terminals of degree 2 in T. When all such non-terminals are deleted from T, our new tree can have at most $2k$ vertices. Note that this tree exactly preserves all flows among terminals.

Now, we crucially observe that in the flow sparsifier H output by Algorithm 2, the capacity between any two terminals x and x' is exactly the probability that x and x' are connected under the random mapping f. We next show that this probability can be computed efficiently.

Let (x, x') be any terminal pair, $\text{lca}(x, x')$ denote their lowest common ancestor in T and r denote the level of $\text{lca}(x, x')$ in T. Moreover, let $V_r^x = \{x, v_{m_x - 1}, \ldots, v_r\}$, $v_r = \text{lca}(x, x')$, be the set of vertices belonging to the shortest path between x and the $\text{lca}(x, x')$. Similarly, define $V_r^{x'} = \{x', v'_{m_{x'} - 1}, \ldots, v_r\}$. Since in Algorithm 2 all non-terminals are mapped independently of each other, we obtain

$$\mathbb{P}[(f(x), f(x')) \in E_H] = 2 \cdot \mathbb{P}[f(v_r) = x] \cdot \mathbb{P}[f(v) = x, \ \forall v \in V_{r-1}^x]$$
$$\cdot \mathbb{P}[f(v') = x', \ \forall v' \in V_{r-1}^{x'}] = \frac{2}{c_r} \cdot \prod_{j=r}^{m_x - 1} \frac{1}{c_j} \prod_{j=r}^{m_{x'} - 1} \frac{1}{c'_j}. \tag{5}$$

where c_j, c'_j are the number of children of the non-terminal v_j, v'_j, respectively.

The above expression suggest that one should build an efficient data-structure for T that answers queries of the form "What is the product of the elements associated with vertices along the path from x to x' in T?". This problem is known as *The Tree Product Query* problem. For an arbitrary tree with n vertices, Alon and Schieber [1] show that in order to answer each Tree Product query in at most $O(\alpha(n))$ steps, an $O(n)$ preprocessing time is sufficient.

Now we are ready to give our deterministic procedure. We first apply our initial preprocessing step in $O(n)$ time. Since the resulting tree has at most $2k$ vertices, it takes $O(k)$ time to preprocess the tree such that every internal vertex knows the number of its children. Next, using $O(k)$ preprocessing, we build a data-structure for the Tree Product Query problem. Now, for every terminal pair (x, x') we can compute in $O(\alpha(2k))$ time the capacity of (x, x') in H from the Tree product query between x and x' and Eq. (5). Since there are at most $O(k^2)$ terminal pairs, we get a running time of $O(n + k^2\alpha(2k))$. The correctness is immediate from the above observations.

Extension to Arbitrary Trees. The above algorithm can be extended to arbitrary trees (deferred to the full version). This leads to the following theorem:

Theorem 1. *Given an unweighted tree* $T = (V, E)$, $K \subset V$, *there exists a 2-quality flow sparsifier* H. *Moreover,* H *can be viewed as a convex combination over connected 0-extensions of* T.

3   Lower Bound

In this section we present a $2 - o(1)$ lower bound on the quality of any cut sparsifier for a star graph. Since previous lower bounds relied on non-planar graph instances, this is the first non-trivial lower bound for arbitrary cut sparsifiers on planar graphs. The result extends to the stronger notion of flow sparsifiers.

The main idea behind our approach is to exploit the symmetries of the star graph. We observe that these symmetries induce other symmetries on the cut structure of the graph. This simplifies the structure of an optimal cut-sparsifier.

Let $G = (K \cup \{v\}, E)$, be an unweighted star with k terminals. Let π' be any permutation of K. We extend π' to a permutation π of $K \cup \{v\}$ by setting $\pi(x) = \pi'(x), \forall x \in K$ and $\pi(v) = v$. Now, for any $U \subset K \cup \{v\}$ and any such a permutation π, we use the symmetry $\mathrm{cap}_G(\delta(U)) = \mathrm{cap}_G(\delta(\pi(U)))$. The latter implies that for any $S \subset K$, $\mathrm{mincut}_G(S, K \setminus S) = \mathrm{mincut}_G(\pi(S), K \setminus \pi(S))$.

For a cut sparsifier H of quality q for G, we show that $\pi(H)$, i.e., the graph obtained by renaming all vertices of H according to permutation π, is also a cut sparsifier of quality q for G. Indeed, for any $S \in K$, $\mathrm{cap}_{\pi(H)}(\delta(S)) = \mathrm{cap}_H(\delta(\pi^{-1}(S))) \geq \mathrm{mincut}_G(\pi^{-1}(S), K \setminus \pi^{-1}(S)) = \mathrm{mincut}_G(S, K \setminus S)$. Symmetrically, one can show that $\mathrm{cap}_{\pi(H)}(\delta(S)) \leq q \cdot \mathrm{mincut}_G(S, K \setminus S)$.

Lemma 8. *A convex combination of any two cut sparsifiers with the same quality gives a new cut sparsifier with the same or better quality.*

Lemma 9. *For the star graph G defined as above, there exists an optimum cut sparsifier H, which is a complete graph with uniform edges-weights.*

Proof. First, we observe by Lemma 8 that if we have two cut sparsifiers with the same quality, taking their convex combination gives a new cut sparsifier with the same or better quality. Suppose we are given some optimum cut sparsifier H'. We can generate $k!$ different cut sparsifiers by considering all possible permutations π as defined above. By the above arguments, for each π, we know that $\pi(H')$ is also an optimum cut sparsifier. Taking the convex combination over $k!$ such sparsifiers, we obtain a complete graph H with uniform edge-weights.   □

Lemma 10. *If H is uniform weighted complete graph that is an optimum cut sparsifier for the star graph G and k even, the edge weight must be at least $2/k$.*

Proof. By definition, H must dominate the terminal cut that has $k/2$ vertices on one side. The minimum value of such a cut in G is $k/2$. The number of edges that cross such a cut in H is $k^2/4$. Since H has uniform edge-weights, this gives that the edge weight must be at least $2/k$.   □

Theorem 2. *Let $G = (K \cup \{v\}, E)$ be an unweighted star with k terminals. Then, there is no cut sparsifier H that achieves quality better than $2 - o(1)$.*

Proof. By the above lemmas, we can assume without loss of generality that H is a complete graph with uniform edge-weights, where this edge weight is at least $2/k$. Hence, a cut that has a singleton terminal vertex on one side has capacity $2(k-1)/k = 2(1 - 1/k)$ in H but it has minimum cut value 1 in G. The latter implies that the quality of H must be at least $2(1 - 1/k)$.   □

4   Improved Results for Quasi-Bipartite Graphs

In this section, we present two new tradeoffs for flow sparsifiers in quasi-bipartite graphs. For this family of graphs, Andoni et al. [3] show how to obtain flow sparsifier with very good quality and moderate size. Specifically, they obtain an $(1 + \varepsilon)$-quality flow sparsifier of size $\widetilde{O}(k^7/\varepsilon^3)$. In the original definition of flow sparsifiers, Leighton and Moitra [14] studied the version where sparsifiers lie only on the terminals, i.e., $V_H = K$. For this restricted setting, we obtain a flow sparsifier of quality 2.

Exact Cut Sparsifier (a.k.a Mimicking Networks) were introduced by Hagerup et al. [9]. In their work they show that general graphs admit exact cut sparsifiers of size doubly exponential in k. As a second result, we show that unit weighted quasi-bipartite graphs admit an exact flow sparsifier of size 2^k.

A graph G with terminals K is *quasi-bipartite* if the non-terminals form an independent set. Throughout this section we assume w.l.o.g. that we are given a bipartite graph with terminals lying on one side and non-terminals in the other (this can achieved by subdividing terminal-terminal edges).

A 2-Quality Flow Sparsifier of Size k. Assume we are given an unweighted bipartite graph G with terminals K. The crucial observation is that we can view G as taking union over stars, where each non-terminal is the center connected to some subset of terminals. Lemma 2 allows us to study these stars independently. Then, for every such star, we apply Lemma 7 to obtain a flow sparsifier only on the terminals belonging to that star. Finally, we merge the resulting sparsifiers and construct a sparsifier H with $V(H) = K$ by another application of Lemma 7. Since the quality of every star in isolation is 2 or better, H is also a 2-quality flow sparsifier.

We note that Lemma 7 only works for unweighted trees. There is an easy extension that gives a similar lemma for weighted stars.

Lemma 11. *Let $G = (K \cup \{u\}, E, c)$ be a weighted star with k terminals. Then G admits a 2-quality flow sparsifier H of size k.*

Applying the decomposition and merging lemma similarly to the unweighted case leads to the following theorem:

Theorem 3. *Let $G = (V, E, c)$ with $K \subset V$ be a weighted quasi-bipartite graph. Then G admits a 2-quality flow sparsifier H of size k.*

An Exact Flow Sparsifier of Size 2^k. In what follows it will be convenient to work with an equivalent definition for Flow Sparsifiers. Let $\lambda_G(d)$ denote the maximum fraction of *concurrent flow* when routing demand d among terminals in graph G. Then $H = (V_H, E_H, c_H)$ with $K \subset V_H$ is a *flow sparsifier* of G with quality $q \geq 1$ if for all demand functions d, $\lambda_G(d) \leq \lambda_H(d) \leq q \cdot \lambda_G(d)$.

The high level idea of our approach is to create "types" for non-terminals and then merge all non-terminals of the same type into a single non-terminal (i.e., add infinity capacity among all non-terminals of the same type). The main difficulty is to define the right types and show that the merging does not affect the multi-commodity flow structure among the terminals. A similar approach was developed by Andoni et al. [3], but their guarantees applies only to approximate flow sparsifier.

We start by defining types. We say that two non-terminals u, v are of the same type if they are incident to the same subset of terminals. Non-terminals of the same type form groups. Note that a non-terminal belongs to an unique group. The *size* of the group is the number of non-terminals belonging to that group. Since the set of non-terminals is an independent set, by Lemma 2, we can construct sparsifiers for each group independently. Our final sparsifier is obtained by merging the sparsifiers over all groups. By another application of Lemma 2, if the sparsifiers of the groups are exact flow sparsifiers, then the final sparsifier is also an exact flow sparsifier for the original graph.

Next, if we replace each group by a single non-terminal, then the size guarantee of the final sparsifier follows from the fact that there are at most 2^k different subsets of terminals. Below we formalize the merging operation within groups.

Let $G_i = (K' \cup \{v_1, \ldots, v_{n_i}\}, E_i, c)$ be a group of size $n_i \geq 2$, where $E_i = \{\{v_j, x\} : j \in \{1, \ldots, n_i\}, x \in K'\}$, $K' \subseteq K$ and $c(e) = 1$, $e \in E_i$. We get:

Lemma 12. *Let G_i with $K' \subset V(G_i)$ be a group of size $n_i \geq 2$ defined as above. Then G_i can be replaced by a star $H_i = (K' \cup \{v_1\}, E_{H_i}, c_{H_i})$ with edge weights $c_{H_i}(e) = n_i$, for all $e \in E_{H_i}$, and which preserves exactly all multicommodity flows between terminals from K'.*

Taking the union over all sparsifiers H_i leads to the following theorem:

Theorem 4. *Let $G = (V, E)$ with $K \subset V$ be a unit weighted quasi-bipartite graph. Then G admits an exact flow sparsifier H of size at most 2^k.*

Proof (Lemma 12). First, observe that we can think of H_i as adding infinity capacity edges between non-terminals in G_i. Then merging into a single non-terminal is done by simply adding edge weights incident to the same terminal. More precisely, let $E_{H_i} = \{(v_r, v_s) : r, s = 1, \ldots, n_i, \ r \neq s\}$. Then, we can assume that $H_i = (K' \cup \{v_1, \ldots v_{n_i}\}, E_i \cup E_{H_i}, c_{H_i})$ where $c_{H_i}(e) = c(e)$ if $e \in E_i$ and $c_{H_i}(e) = \infty$ if $e \in E_{H_i}$.

Since we can route every feasible demand from G_i in H_i even without using the infinity-capacity edges, it is immediate that for any demand function d, $\lambda_{H_i}(d) \geq \lambda_{G_i}(d)$. Thus, we only need to show that $\lambda_{H_i}(d) \leq \lambda_{G_i}(d)$. To achieve this, we will use the dual to the maximum concurrent flow problem (i.e., the Fractional Sparsest Cut Problem). The dual problem is the following[2]:

$$
\begin{aligned}
\min \quad & \sum_{j=1}^{n_i} \sum_{x \in K'} \ell_{v_j x} \\
\text{s. t.} \quad & \ell_{s v_j} + \ell_{v_j t} \geq \delta_{st} && \forall \{s, t\} \in \binom{K'}{2}, \ \forall j \in \{1, \ldots, n_i\} \\
& \sum_{\{s,t\} \in \binom{K'}{2}} d_{st} \delta_{st} \geq 1 \\
& \ell_e \geq 0, \quad \delta_{st} \geq 0.
\end{aligned}
\tag{6}
$$

---

[2] Note that the dual requires that δ_{st} is at most the length of the shortest s-t path. In our scenario this is always a 2-hop path. Hence, the above formulation is correct.

Let d be an arbitrary demand function. Moreover, let $\{\ell_e, \delta_{st}\}$ be an optimal solution of value $\lambda_{G_i}(d)$ for the LP in Eq. (6), where δ_{st} is the shortest-path distance induced by the length assignment ℓ. We first modify this solution and get a new feasible solution with the same cost and a certain structure that we will later exploit.

The modification works as follows. For every terminal we create a set of edges incident to that terminal. Then, within each set, we replace the length of each edge by the total *average* length of the group. Specifically, for every $x \in K'$, let $E_x = \{(v_j, x) : j = 1, \ldots, n_i\}$ be the set of edges incident to x.

The new edge lengths are defined as follows: $\widetilde{\ell}_{v_j x} = \sum_{e \in E_x} \ell_e / n_i, \forall x \in K', \forall j = 1, \ldots, n_i$. Let $\widetilde{\delta}_{st}$ be the new shortest-path distance induced by the length assignment $\widetilde{\ell}$. In order for $\{\widetilde{\ell}_e, \widetilde{\delta}_{st}\}$ to be feasible, we need to show that $\widetilde{\delta}$ dominates δ, i.e., $\widetilde{\delta}_{st} \geq \delta_{st}$, for every pair $s, t \in K'$. Indeed, since edge lengths within groups are the same, we get that for every pair $s, t \in K'$:

$$\widetilde{\delta}_{st} = \widetilde{\ell}_{sv_1} + \widetilde{\ell}_{v_1 t} = \frac{1}{n_i} \sum_{e \in E_s} \ell_e + \frac{1}{n_i} \sum_{e \in E_t} \ell_e = \frac{1}{n_i} \sum_{j=1}^{n_i} \left(\ell_{sv_j} + \ell_{v_j t} \right)$$

$$\geq \min_{j \in \{1, \ldots, n_i\}} \{\ell_{sv_j} + \ell_{v_j t}\} \geq \delta_{st}.$$

Additionally, observe that the new solution has the same optimal value, namely $\lambda_{G_i'}^*(d) = \sum_{j=1}^{n_i} \sum_{x \in K'} \ell_{v_j x} = \sum_{j=1}^{n_i} \sum_{x \in K'} \widetilde{\ell}_{v_j x}$. Hence, we can assume without loss of generality that an optimal solution satsifies: $\widetilde{\ell}_{v_1 x} = \ldots = \widetilde{\ell}_{v_{n_i} x}, \forall x \in K'$. Now, we add edges (v_i, v_j) to G_i and set $\widetilde{\ell}_{v_i v_j} = 0$, for all $i, j = 1, \ldots, n_i$. Note that shortest-path distances $\widetilde{\delta}_{st}$ do not change by this modification. Therefore, by adding these zero edge lengths between the non-terminals, we still get an optimum solution $\{\widetilde{\ell}_e, \widetilde{\delta}_{st}\}$ for the LP in (6).

Finally, let us define the dual problem for the star H_i:

$$\begin{aligned}
\min \quad & \sum_{j=1}^{n_i} \sum_{x \in K'} \ell_{v_j x} \\
\text{s. t.} \quad & \sum_{e \in P_{st}} \ell_e \geq \delta_{st} \qquad \forall \{s,t\} \in \binom{K'}{2}, \forall s\text{-}t \text{ paths on } E \cup E_{H_i} \\
& \sum_{\{s,t\} \in \binom{K'}{2}} d_{st} \delta_{st} \geq 1 \\
& \ell_e \geq 0, \quad \delta_{st} \geq 0, \quad \forall e \in E_{H_i} \, \ell_e = 0.
\end{aligned} \qquad (7)$$

It follows from above that $\{\widetilde{\ell}_e, \widetilde{\delta}_{st}\}$ is a feasible solution for the LP in (7). Hence, $\lambda_{H_i}(d) \leq \lambda_{G_i}(d)$, what we were after.   □

References

1. Alon, N., Schieber, B.: Optimal preprocessing for answering on-line product queries. Technical report, Tel Aviv University (1987)
2. Andersen, R., Feige, U.: Interchanging distance and capacity in probabilistic mappings. CoRR, arXiv:abs/0907.3631 (2009)
3. Andoni, A., Gupta, A., Krauthgamer, R.: Towards $(1+\varepsilon)$-approximate flow sparsifiers. In: Proceedings of the 25th SODA, pp. 279–293 (2014)

4. Benczúr, A.A., Karger, D.R.: Approximating s-t minimum cuts in $\tilde{O}(n^2)$ time. In: Proceedings of the 28th STOC, pp. 47–55 (1996)
5. Chekuri, C., Shepherd, F.B., Oriolo, G., Scutellà, M.G.: Hardness of robust network design. Networks **50**(1), 50–54 (2007)
6. Cheung, Y.K., Goranci, G., Henzinger, M.: Graph minors for preserving terminal distances approximately - lower and upper bounds. In: Proceedings of the 43rd ICALP, pp. 131:1–131:14 (2016)
7. Chuzhoy, J.: On vertex sparsifiers with steiner nodes. In: Proceedings of the 44th STOC, pp. 673–688 (2012)
8. Gupta, A.: Steiner points in tree metrics don't (really) help. In: Proceedings of the 12th SODA, pp. 220–227 (2001)
9. Hagerup, T., Katajainen, J., Nishimura, N., Ragde, P.: Characterizing multiterminal flow networks and computing flows in networks of small treewidth. J. Comput. Syst. Sci. **57**(3), 366–375 (1998)
10. Kamma, L., Krauthgamer, R., Nguyen, H.L.: Cutting corners cheaply, or how to remove steiner points. SIAM J. Comput. **44**(4), 975–995 (2015)
11. Khan, A., Raghavendra, P.: On mimicking networks representing minimum terminal cuts. Inf. Process. Lett. **114**(7), 365–371 (2014)
12. Krauthgamer, R., Nguyen, H.L., Zondiner, T.: Preserving terminal distances using minors. SIAM J. Discrete Math. **28**(1), 127–141 (2014)
13. Krauthgamer, R., Rika, I.: Mimicking networks and succinct representations of terminal cuts. In: Proceedings of the 24th SODA, pp. 1789–1799 (2013)
14. Leighton, F.T., Moitra, A.: Extensions and limits to vertex sparsification. In: Proceedings of the 42nd STOC, pp. 47–56 (2010)
15. Makarychev, K., Makarychev, Y.: Metric extension operators, vertex sparsifiers and lipschitz extendability. In: Proceedings of the 51th FOCS, pp. 255–264 (2010)
16. Moitra, A.: Approximation algorithms for multicommodity-type problems with guarantees independent of the graph size. In: Proceedings of the 50th FOCS (2009)
17. Räcke, H.: Optimal hierarchical decompositions for congestion minimization in networks. In: Proceedings of the 40th STOC, pp. 255–264 (2008)
18. Räcke, H., Shah, C., Täubig, H.: Computing cut-based hierarchical decompositions in almost linear time. In: Proceedings of the 25th SODA, pp. 227–238 (2014)
19. Spielman, D.A., Teng, S.: Spectral sparsification of graphs. SIAM J. Comput. **40**(4), 981–1025 (2011)

Scenario Submodular Cover

Nathaniel Grammel[1], Lisa Hellerstein[1(✉)], Devorah Kletenik[2], and Patrick Lin[3]

[1] Department of Computer Science and Engineering,
NYU Tandon School of Engineering, Brooklyn, NY, USA
{ngrammel,lisa.hellerstein}@nyu.edu
[2] Department of Computer and Information Science, Brooklyn College,
City University of New York, New York, NY, USA
kletenik@sci.brooklyn.cuny.edu
[3] Department of Computer Science, University of Illinois at Urbana-Champaign,
Champaign, IL, USA
plin15@illinois.edu

Abstract. We introduce the Scenario Submodular Cover problem. In this problem, the goal is to produce a cover with minimum expected cost, with respect to an empirical joint probability distribution, given as input by a weighted sample of realizations. The problem is a counterpart to the Stochastic Submodular Cover problem studied by Golovin and Krause [6], which assumes independent variables. We give two approximation algorithms for Scenario Submodular Cover. Assuming an integer-valued utility function and integer weights, the first achieves an approximation factor of $O(\log Qm)$, where m is the sample size and Q is the goal utility. The second, simpler algorithm achieves an approximation factor of $O(\log QW)$, where W is the sum of the weights. We achieve our bounds by building on previous related work (in [4,6,15]) and by exploiting a technique we call the Scenario-OR modification. We apply these algorithms to a new problem, Scenario Boolean Function Evaluation. Our results have applciations to other problems involving distributions that are explicitly specified by their support.

1 Introduction

The Submodular Cover problem is a fundamental problem in submodular optimization that generalizes the classical NP-complete Set Cover problem. Adaptive versions of this problem have applications to a number of other problems, notably machine learning problems where the goal is to build a decision tree or strategy of minimum expected cost. Examples of such problems include entity identification (exact learning with membership queries), classification (equivalence class determination), and decision region identification (cf. [1,6,7,11]). Other applications include reducing expected prediction costs for learned Boolean classifiers, given attribute costs [5].

Previous work on *Stochastic* Submodular Cover assumes independence of the variables of the probability distribution. Optimization is performed with respect to this distribution. We consider a new version of the problem that we call

© Springer International Publishing AG 2017
K. Jansen and M. Mastrolilli (Eds.): WAOA 2016, LNCS 10138, pp. 116–128, 2017.
DOI: 10.1007/978-3-319-51741-4_10

Scenario Submodular Cover, that removes the independence assumption. In this problem, optimization is with respect to an input distribution given explicitly by its support (with associated probability weights). We give approximation algorithms solving Scenario Submodular Cover over discrete distributions.

In generic terms, an adaptive submodular cover problem is a sequential decision problem where we must choose items one by one from an item set $N = \{1, \ldots, n\}$. Each item has an initially unknown state, which is a member of a finite state set Γ. The state of an item is revealed only after we have chosen the item. We represent a subset $S \subseteq N$ of items and their states by a vector $x \in (\Gamma \cup \{*\})^n$ where $x_i = *$ if $i \notin S$, and x_i is the state of item i otherwise. We are given a monotone, submodular[1] utility function $g \colon (\Gamma \cup \{*\})^n \to \mathbb{Z}_{\geq 0}$. It assigns a non-negative integer value to a subset of the items where the value can depend on the states of the items. There is a non-negative goal utility value Q, such that $g(a) = Q$ for all $a \in \Gamma^n$. There is a cost associated with choosing each item, which we are given. In distributional settings, we are also given the joint distribution of the item states. We continue choosing items until their utility value is equal to the goal utility, Q. The problem is to determine the adaptive order in which to choose items so as to minimize expected cost (in distributional settings) or worst-case cost (in adversarial settings).

Stochastic Submodular Cover is an adaptive submodular cover problem in a distributional setting. In this problem, the state of each item is an independent random variable. The distributions of the variables are given as input. Golovin and Krause introduced a simple algorithm for this problem, called Adaptive Greedy, achieving an approximation factor of $O(\log Q)$. Another algorithm for the problem, called Adaptive Dual Greedy, was presented by Deshpande et al. [5]. These algorithms have been useful in solving other stochastic optimization problems, which can be reduced to Stochastic Submodular Cover through the construction of appropriate utility functions (e.g., [2,5,7,11]).

The problem we study, *Scenario Submodular Cover* (Scenario SC), is also a distributional, adaptive submodular cover problem. The distribution is given by a weighted sample. Each element of the sample is a vector in Γ^n, representing an assignment of states to the items in N. Associated with each assignment is a positive integer weight. The sample and its weights define a joint distribution on Γ^n, where the probability of a vector γ in the sample is proportional to its weight. (The probability of a vector in Γ^n that is not in the sample is 0.) As in Stochastic Submodular Cover, the problem is to choose the items and achieve utility Q, while minimizing expected cost. However, because proofs of results for the Stochastic Submodular Cover problem typically rely on the independence assumption, they do not apply to the Scenario SC problem.

Results. We present *Mixed Greedy*, an approximation algorithm for the Scenario SC problem that uses two different greedy criteria. It is a generalization of the

---

[1] The definitions "monotone" and "submodular," for state-dependent utility functions, has not been standardized. We define these terms in Sect. 2. In the terminology used by Golovin and Krause [6], g is *pointwise* monotone and pointwise submodular.

algorithm of Cicalese et al. [4] for Equivalence Class Determination (also called Group Identification and Discrete Function Evaluation). Our analysis uses the the same basic approach as that used by Cicalese et al., but the proof of their main technical lemma does not apply to our problem. We replace it with a histogram proof similar to that used in [15] for Min-Sum Submodular Cover.

The approximation factor achieved by Mixed Greedy for Scenario SC is $O(\frac{1}{\rho} \log Q)$, where ρ is a technical quantity associated with utility function g. The utility function constructed for the Equivalence Class Determination problem has constant ρ, but this is not the case in general.

To achieve a better bound for other problems, we present a modified version of Mixed Greedy, which uses an existing construction in a novel way. The existing construction produces the OR of two monotone submodular functions with goal values (cf. [5,9]). We apply this construction to g and to another utility function based on the sample, to get a new monotone, submodular function g_S, for which ρ is constant. We call the transformation of g and the sample into g_S the *Scenario-OR* modification.

Once g_S is constructed, Mixed Greedy is run on g_S with goal value Qm, where m is the size of the sample. We show that the resulting algorithm, *Scenario Mixed Greedy*, achieves an $O(\log Qm)$ approximation factor for any Scenario SC problem.

In addition to Mixed Greedy, we also present a simpler, more efficient algorithm for the Scenario SC problem, *Scenario Adaptive Greedy*, with a worse approximation bound. It is based on the Adaptive Greedy algorithm of Golovin and Krause. However, the approximation bound proved by Golovin and Krause [6] for Adaptive Greedy depends on the assumption that g and the distribution defined by the sample weights jointly satisfy *adaptive submodularity*. This is not the case for general Scenario SC instances. Scenario Adaptive Greedy is obtained by modifying Adaptive Greedy using a weighted version of the Scenario-OR modification. Scenario Adaptive Greedy combines g and the weighted sample to obtain a modified utility function g_W, having goal utility QW. Scenario Adaptive Greedy then applies Adaptive Greedy to g_W. We prove that g_W and the distribution defined by the weights jointly satisfy adaptive submodularity. Using the existing approximation bound for Adaptive Greedy then implies a bound of $O(\log QW)$ for Scenario Adaptive Greedy, where W is the sum of the weights.

The constructions of g_S and g_W are similar to constructions in work on Equivalence Class Determination and related problems (cf. [1–3,7]). Our proof of adaptive submodularity uses the approach of showing that a certain function is non-decreasing along a path between two points. This approach was used before (cf. [2,3,7]) but our problem is more general and our proof differs.

Previously, applying ordinary Adaptive Greedy to solve sample-based problems required constructing a utility function g, and then proving adaptive submodularity of g and the distribution on the weighted sample. The proof could be non-trivial (see, e.g., [1,3,7,11]). With our approach, one can get an approximation bound with Adaptive Greedy by proving only submodularity of g, rather than adaptive submodularity of g and the distribution. Proofs of submodularity are generally easier. Also, the OR construction used in Sect. 2 preserves submodularity, but not Adaptive Submodularity [2].

Given monotone, submodular g with goal value Q, we can use our algorithms to obtain three approximation results for the associated Scenario SC problem: $O(\frac{1}{\rho}\log Q)$ with Mixed Greedy, $O(\log Qm)$ with Scenario Mixed Greedy, and $O(\log QW)$ with Scenario Adaptive Greedy.

Assuming the costs c_i are integers, and letting $C = \sum_i c_i$, we note that applying the "Kosaraju trick" (first used by Kosaraju et al. in [13]) to Scenario Adaptive Greedy yields a bound of $O(\log QmC)$ instead of $O(\log QW)$. See [6] for a similar use of the trick.

After the appearance of a preliminary version of this paper [8], Navidi et al. [14] presented a new algorithm solving a generalization of the Scenario SC problem. It achieves the $O(\log Qm)$ bound of Scenario Mixed Greedy using a single greedy rule, different from the one used in Scenario Adaptive Greedy. Their algorithm can be applied to problems where there is a distinct monotone submodular function for each scenario.

Applications. The Scenario SC problem has many applications. As an example, consider the query learning problem of identifying an unknown hypothesis h from a hypothesis class $\{h_1, \ldots, h_m\}$ by asking queries from the set $\{q_1, \ldots, q_n\}$. The answer to each query is 0 or 1, and we are given an $m \times n$ table D where $D[i, j]$ is the answer to q_i for h_j. Each pair of hypotheses differs on at least one query. Suppose there is a given cost c_i for asking query q_i, and each h_j has a given prior probability p_j. The problem is to build a decision tree (querying procedure) for identifying h, minimizing expected query cost, assuming h is drawn with respect to the p_j. View the q_i as items $i \in N$, the h_j as scenarios, and the answer to q_i as the state of item i. Represent answers to queries asked so far as a partial assignment $b \in \{0, 1, *\}^n$ where $b_i = *$ means q_i has not been asked. Define utility function $g \colon \{0, 1, *\}^n \to \mathbb{Z}_{\geq 0}$ whose value on $b \in \{0, 1, *\}^n$ is $\min\{m - 1, r(b)\}$ where $r(b) = |\{h_j \mid \exists i \text{ such that } b_i \neq * \text{ and } D[i, j] \neq b_i\}|$. Function g is monotone and submodular. Further, $g(b) = m - 1$ iff the answers in b uniquely identify h. Building a decision tree with minimum expected decision cost is equivalent to solving Scenario SC for g with goal value $Q = m - 1$, for costs c_i and weights proportional to the p_i. An algorithm with an approximation bound of $O(\log m)$ for this problem was first presented by [10].

Equivalence Class Determination is a generalization of the query learning problem where in addition to D, we are given a partition of the h_j into equivalence classes. The decision tree must just identify the class to which h belongs. This problem can also be seen as a Scenario SC problem, using the "Pairs" utility function of Cicalese et al., which has goal value $Q = O(m^2)$ [4]. Applying our Scenario Mixed Greedy bound to this utility function yields an approximation bound of $O(\log m)$, matching the bound of Cicalese et al.

Our bound on Scenario Mixed Greedy yields a new approximation bound for the Decision Region Identification problem studied by Javdani et al. [11], which is an extension of Equivalence Class Determination. They define a utility function whose value is a weighted sum of hyperedges cut in a certain hypergraph. We define a utility function whose value is the *number* of hyperedges cut. Using Mixed Greedy with this function yields an approximation bound of $O(k \log m)$, where k is a parameter associated with the problem, and m is the sample size. In contrast, the bound in [11] is $O(k \log(\frac{W}{w_{min}}))$, where w_{min} is the minimum weight on a realization in the sample. (The recent paper of Navidi et al. [14] gives a further bound.)

We can apply our algorithms to Scenario BFE (Boolean Function Evaluation) problems, which we introduce here. These problems are a counterpart to the Stochastic BFE problems[2] studied in AI, operations research, and in learning with attribute costs (see e.g., [5,12,16]). In a Scenario BFE problem, we are given a representation of a Boolean function $f\colon \{0,1\}^n \to \{0,1\}$. For each $i \in \{1,\ldots,n\}$, we are given $c_i > 0$, the cost of obtaining the value of the ith bit of an initially unknown $a \in \{0,1\}^n$. We are given a weighted sample $S \subseteq \{0,1\}^n$. The problem is to compute a (possibly implicit) decision tree computing f, minimizing the expected cost of evaluating f on $a \in \{0,1\}^n$ using the tree. The expectation is with respect to the distribution defined by the sample weights.

Deshpande et al. [5] gave approximation algorithms for some Stochastic BFE problems that work by constructing a monotone, submodular utility function g and running Adaptive Greedy. By substituting the sample-based algorithms in this paper in place of Adaptive Greedy, we obtain results for analogous Scenario BFE problems. For example, using Mixed Greedy, we obtain an $O(k \log n)$ approximation for the Scenario BFE problem for k-of-n functions, a bound that is independent of sample size. Details are in the full version of the paper.

We note that the Scenario BFE problem differs from the function evaluation problem considered by Cicalese et al. [4]. In that problem, the decision tree must only compute f correctly on assignments $a \in \{0,1\}^n$ in the sample, while in Scenario BFE the tree must compute f correctly on all $a \in \{0,1\}^n$. Also, in Scenario BFE we assume function f is given with the sample, and we consider particular types of functions f.

2   Definitions

Let $N = \{1,\ldots,n\}$ be the set of *items* and Γ be a finite set of states. A *sample* is a subset of Γ^n. A *realization* is an element $a \in \Gamma^n$, representing an assignment of states to items, where for $i \in N$, a_i represents the state of item i. We also refer to an element of Γ^n as an *assignment*.

We call $b \in (\Gamma \cup \{*\})^n$ a *partial* realization. Partial realization b represents the subset $I = \{i \mid b_i \neq *\}$ where each item $i \in I$ has state b_i. For $\gamma \in \Gamma$, the quantity $b_{i \leftarrow \gamma}$ denotes the partial realization produced from b by setting $b_i = \gamma$. For $b, b' \in (\Gamma \cup \{*\})^n$, b' is an *extension* of b, written $b' \succeq b$, if $b'_i = b_i$ for all $b_i \neq *$. We use $b' \succ b$ to denote that $b' \succeq b$ and $b' \neq b$.

---

[2] In the Operations Research literature, Stochastic Function Evaluation is often called Sequential Testing or Sequential Diagnosis.

Let $g\colon (\Gamma \cup \{*\})^n \to \mathbb{Z}_{\geq 0}$ be a utility function. Function $g\colon (\Gamma \cup \{*\})^n \to \mathbb{Z}_{\geq 0}$ has *goal value* Q if $g(a) = Q$ for all realizations $a \in \Gamma^n$. We define $\Delta g(b, i, \gamma) := g(b_{i \leftarrow \gamma}) - g(b)$.

A standard utility function is a set function $f\colon 2^N \to \mathbb{R}_{\geq 0}$. It is monotone if for all $S \subset S' \subseteq N$, $f(S) \leq f(S')$. It is submodular if in addition, for $i \in N - S$, $f(S \cup \{i\}) - f(S) \geq f(S' \cup \{i\}) - f(S')$. We extend definitions of monotonicity and submodularity to (state-dependent) function $g\colon (\Gamma \cup \{*\})^n \to \mathbb{Z}_{\geq 0}$ as follows:

- g is *monotone* if for $b \in (\Gamma \cup \{*\})^n$, $i \in N$ such that $b_i = *$, and $\gamma \in \Gamma$, we have $g(b) \leq g(b_{i \leftarrow \gamma})$
- g is *submodular* if for all $b, b' \in (\Gamma \cup \{*\})^n$ such that $b' \succ b$, $i \in N$ such that $b_i = b'_i = *$, and $\gamma \in \Gamma$, we have $\Delta g(b, i, \gamma) \geq \Delta g(b'i, \gamma)$.

Let $\mathcal{D}$ be a probability distribution on Γ^n. Let X be a random variable drawn from $\mathcal{D}$. For $a \in \Gamma^n$ and $b \in (\Gamma \cup \{*\})^n$, we define $\Pr[a \mid b] := \Pr[X = a \mid a \succeq b]$. For i such that $b_i = *$, we define $\mathbb{E}[\Delta g(b, i, \gamma)] := \sum_{a \in \Gamma^n : a \succeq b} \Delta g(b, i, a_i) \Pr[a \mid b]$.

- g is *adaptive submodular with respect to* $\mathcal{D}$ if for all b', b such that $b' \succ b$, $i \in N$ such that $b_i = b'_i = *$, and $\gamma \in \Gamma$, we have $\mathbb{E}[\Delta g(b, i, \gamma)] \geq \mathbb{E}[\Delta g(b', i, \gamma)]$.

Intuitively, we can view b as partial information about states of items i in a random realization $a \in \Gamma^n$, with $b_i = *$ meaning the state of item i is unknown. Then g measures the utility of that information, and $\mathbb{E}[\Delta g(b, i, \gamma)]$ is the expected increase in utility that would result from discovering the state of i.

For $g\colon (\Gamma \cup \{*\})^n \to \mathbb{Z}_{\geq 0}$ with goal value Q, and $b \in (\Gamma \cup \{*\})^n$ and $i \in N$, where $b_i = *$, let $\gamma_{b,i}$ be the state $\gamma \in \Gamma$ such that $\Delta g(b, i, \gamma)$ is minimized (if more than one exists, choose one arbitrarily). Thus $\gamma_{b,i}$ is the state of item i that would produce the smallest increase in utility, and thus is "worst-case" in terms of utility gain, if we start from b and then discover the state of i.

For fixed $g\colon (\Gamma \cup \{*\})^n \to \mathbb{Z}_{\geq 0}$ with goal value Q, we define an associated quantity ρ, as follows:

$$\rho := \min \frac{\Delta g(b, i, \gamma)}{Q - g(b)} \tag{1}$$

where the minimization is over b, i, γ, where $b \in (\Gamma \cup \{*\})^n$ such that $g(b) < Q$, $i \in N$ such that $b_i = *$, and $\gamma \in \Gamma - \{\gamma_{b,i}\}$.

Intuitively, when the state of item i is discovered, the distance between the utility achieved and the goal utility is reduced by some fraction (possibly zero). The fraction can vary depending on item state. Parameter ρ equals the smallest possible value for the fraction associated with the next-to-worst case state, starting from any partial realization, and considering any item i whose state is about to be discovered.

An instance of the Scenario SC problem is a tuple (g, Q, S, w, c), where $g\colon (\Gamma \cup \{*\})^n \to \mathbb{Z}_{\geq 0}$ is an integer-valued, monotone submodular utility function with goal value $Q > 0$, $S \subseteq \Gamma^n$, $w : S \to \mathbb{Z}_{\geq 0}^n$ assigns a weight to each realization $a \in S$, and $c \in \mathbb{R}_{>0}^n$ is a *cost vector*. We consider a setting where we select items without repetition from the set of items N, and the states of the items correspond to an initially unknown realization $a \in \Gamma^n$. Each time we select an item, the state a_i of the item is revealed. The selection of items can be adaptive, in that the next item chosen can depend on the states of the previous items. We continue to choose items until $g(b) = Q$, where b is the partial realization representing the states of the chosen items.

The Scenario SC problem asks for an adaptive order in which to choose the items (i.e. a *strategy*), until goal value Q is achieved, such that the expected sum of costs of the chosen items is minimized. The expectation is with respect to the distribution on Γ^n that is proportional to the weights on the assignments in the sample: $\Pr[a] = 0$ if $a \notin S$, and $\Pr[a] = \frac{w(a)}{W}$ otherwise, where $W = \sum_{a \in S} w(a)$. We call this the *sample distribution* defined by S and w and denote it by $\mathcal{D}_{S,w}$.

The strategy corresponds to a decision tree. The internal nodes of the tree are labeled with items $i \in N$, and each such node has $|\Gamma|$ children, one for each state $\gamma \in \Gamma$. We refer to the child corresponding to state γ as the γ-child. Each root-leaf path in the tree is associated with a partial realization b such that for each consecutive pairs of nodes v and v' on the path, if i is the label of v, and v' is the γ-*child* of v, then $b_i = \gamma$. If i does not label any node in the path, then $b_i = *$. The tree may be output in an implicit form (for example, in terms of a greedy rule), specifying how to determine the next item to choose, given the previous items chosen and their states. Although realizations $a \notin S$ do not contribute to the expected cost of the strategy, we require the strategy to achieve goal value Q on *all* realizations $a \in \Gamma^n$.

We will use an existing "OR construction," a method for taking the OR of two utility functions [5,9]. It is a method for combining two monotone submodular utility functions g_1 and g_2 defined on $(\Gamma \cup \{*\})^n$, and values Q_1 and Q_2, into a new monotone submodular utility function g. For $b \in (\Gamma \cup \{*\})^n$,

$$g(b) = Q_1 Q_2 - (Q_1 - g_1(b))(Q_2 - g_2(b)) \tag{2}$$

If for all $a \in \Gamma^n$, $g_1(a) = Q_1$ or $g_2(a) = Q_2$, then $g(a) = Q_1 Q_2$ for all $a \in \Gamma^n$.

3 Mixed Greedy

Mixed Greedy is a generalization of the approximation algorithm developed by Cicalese et al. for the Equivalence Class Determination problem [4]. That algorithm solves the Scenario Submodular Cover problem for a particular "Pairs" utility function associated with Equivalence Class Determination. In contrast, Mixed Greedy can be used on any monotone, submodular utility function g. Following Cicalese et al., we present Mixed Greedy as outputting a tree. If the strategy is only to be used on one realization, it is not necessary to build the entire tree.

3.1 Algorithm

The Mixed Greedy algorithm builds a decision tree for a Scenario SC instance (g, Q, S, w, c). The tree is built top-down, and is structured as described at the end of Sect. 2.

Algorithm 1

Procedure MixedGreedy(g, Q, S, w, c, b)

1: **If** $g(b) = Q$ **then return** a single (unlabeled) leaf l
2: Let T be an empty tree
3: $N' \leftarrow \{i : b_i = *\}$
4: For $i \in N'$, $\sigma_i \leftarrow \arg\min_{\gamma \in \Gamma} \Delta g(b, i, \gamma)$
5: Define $g' : 2^{N'} \to \mathbb{Z}_{\geq 0}$ such that for all $U \subseteq N'$, $g'(U) = g(b_U) - g(b)$, where b_U is the extension of b produced by setting $b_i = \sigma_i$ for all $i \in U$.
6: $B \leftarrow$ FindBudget(N', g', c), $spent \leftarrow 0$, $spent_2 \leftarrow 0$, $k \leftarrow 1$
7: $I \leftarrow \{i \in N' | c_i \leq B\}$
8: For all $R \subseteq I$, define $D_R := \{a \in S | a \succeq b$ and $a_i \neq \sigma_i$ for some $i \in R\}$
9: Define $h : 2^I \to \mathbb{Z}_{\geq 0}$ such that for all $R \subseteq I$, $h(R) = \sum_{a \in D_R} w(a)$
10: $R \leftarrow \emptyset$
11: **repeat**
12:     Let i be an item which maximizes $\frac{h(R \cup \{i\}) - h(R)}{c_i}$ among all items $i \in I$
13:     Let t_k be a new node labeled with item i
14:     **If** $k = 1$ **then** make t_1 the root of T
15:     **else** make t_k the σ_j-child of t_{k-1}
16:     $j \leftarrow i$
17:     **for** every $\gamma \in \Gamma$ such that $\gamma \neq \sigma_i$ **do**
18:         $T^\gamma \leftarrow$ MixedGreedy($g, Q, S, w, c, b_{i \leftarrow \gamma}$)
19:         Attach T^γ to T by making the root of T^γ the γ-child of t_k
20:     $b_i \leftarrow \sigma_i$, $R \leftarrow R \cup \{i\}$, $I \leftarrow I - \{i\}$, $spent \leftarrow spent + c_i$, $k \leftarrow k + 1$
21: **until** $spent \geq B$
22: **repeat**
23:     Let i be an item which maximizes $\frac{\Delta g(b, i, \sigma_i)}{c_i}$ among all items $i \in I$
24:     Let t_k be a node labeled with item i
25:     Make t_k the σ_j-child of t_{k-1}
26:     $j \leftarrow i$
27:     **for** every $\gamma \in \Gamma$ such that $\gamma \neq \sigma_i$ **do**
28:         $T^\gamma \leftarrow$ MixedGreedy($g, Q, S, w, c, b_{i \leftarrow \gamma}$)
29:         Attach T^γ to T by making the root of T^γ the γ-child of t_k
30:     $b_i \leftarrow \sigma_i$, $I \leftarrow I - \{i\}$, $spent_2 \leftarrow spent_2 + c_i$, $k \leftarrow k + 1$
31: **until** $spent_2 \geq B$ **or** $I = \emptyset$
32: $T' \leftarrow$ MixedGreedy(g, Q, S, w, c, b); Attach T' to T by making the root of T' the σ_j-child of t_{k-1}
33: **Return** T

Mixed Greedy works by calling recursive function MixedGreedy, which we present in Algorithm 1. In the initial call, $b = (*, \ldots, *)$. Only the value of parameter b changes between recursive calls. Each call constructs a subtree of the full tree for g, rooted at a node v of that tree. In the call building the subtree rooted at v, b is the partial realization corresponding to the path from the root to v in the full tree: $b_i = \gamma$ if the path includes a node labeled i and its γ-child, and $b_i = *$ otherwise.

The algorithm of Cicalese et al. [4] is essentially the same as Mixed Greedy in the special case where g is equal to their "Pairs" function. Like their algorithm, Mixed Greedy uses a subroutine, FindBudget, that relies on a greedy algorithm of Wolsey for Budgeted Submodular Cover [17]. `FindBudget` is presented in the full version [8] of this paper and is omitted here due to space constraints.

If $g(b) = Q$, then `MixedGreedy` returns an (unlabeled) single node, which will be a leaf of the full tree for g. Otherwise, `MixedGreedy` constructs a tree T. It does so by computing a special realization called σ, and then iteratively using σ to construct a path descending from the root of this subtree, which is called the *backbone*. It uses recursive calls to build the subtrees "hanging" off the backbone. The backbone has a special property: for each node v' in the path, the successor node in the path is the σ_i-child of v', where i is the item labeling node v'.

The backbone is constructed as follows. Using `FindBudget`, `MixedGreedy` computes a lower bound B on the minimum cost required to achieve a fraction of approximately $\frac{1}{3}$ of the goal value Q, assuming we start with partial realization b (Step 6).

After calculating B, `MixedGreedy` constructs the backbone in two stages, using a different greedy criterion in each to determine which item i to place in the current node. In the first stage, corresponding to the first repeat loop, the goal is to remove weight (probability mass) from the backbone, as cheaply as possible. That is, consider an $a \in \Gamma^n$ to be removed from the backbone (or "covered") if i labels a node in the backbone and $a_i \neq \sigma_i$; removing a from the backbone results in the loss of weight $w(a)$ from the backbone. The greedy choice used in the first stage in Step 12 follows the rule of maximizing *bang-for-the-buck*: the algorithm chooses i such that the amount of weight removed from the backbone, divided by c_i, is maximized. In making this choice, it only considers items that have cost at most B. The first stage ends as soon as the total cost of the items in the chosen sequence is at least B. For each item i chosen during the stage, b_i is set to σ_i.

In the second stage, corresponding to the second repeat loop, the goal is to increase utility as measured by g, under the assumption that we already have b, and that the state of each remaining item i is σ_i. The algorithm again uses a bang-for-the-buck rule, choosing the i that maximizes the increase in utility, divided by c_i (Step 23). In making this choice, it again considers only items with cost at most B. The stage ends when the total cost of the items in the chosen sequence is at least B. For each item i chosen during the stage, b_i is set to σ_i.

In Sect. 2, we defined ρ. The way B is chosen guarantees that the updates to b during the two greedy stages cause the value of $Q - g(b)$ to shrink by at least a fraction $\min\{\rho, \frac{1}{9}\}$ before each recursive call. We use this fact to prove the following theorem. The proof can be found in the full version of the paper [8].

Theorem 1. *Mixed Greedy is an approximation algorithm for Scenario Submodular Cover that achieves an approximation factor of $O(\frac{1}{\rho} \log Q)$.*

4   Scenario Mixed Greedy

We now use the Scenario-OR modification to obtain a modified version of Mixed Greedy, called Scenario Mixed Greedy, that eliminates the dependence on ρ in the approximation bound in favor of a dependence on m, the size of the sample. Rather than running Mixed Greedy with g, it first combines g and the sample to produce a new utility function g_S, and then runs Mixed Greedy with g_S, rather than with g. Utility function g_S is produced by combining g with another utility function h_S, using the OR construction described at the end of Sect. 2. Here $h_S \colon (\Gamma \cup \{*\})^n \to \mathbb{Z}_{\geq 0}$, where $h_S(b) = m - |\{a \in S : a \succeq b\}|$ and $m = |S|$. Thus $h_S(b)$ is the total number of assignments that have been eliminated from S because they are incompatible with the partial state information in b. Utility m for h_S is achieved when all assignments in S have been eliminated. Clearly, h_S is monotone and submodular.

When the OR construction is applied to combine g and h_S, the resulting utility function g_S reaches its goal value Qm when all possible realizations of the sample have been eliminated or when goal utility is achieved for g.

In an on-line setting, Scenario Mixed Greedy uses the following procedure to determine the sequence of items to choose on an initially unknown a. We note that the third step in the procedure is present because goal utility Q must be reached even for realizations a not in S.

Scenario Mixed Greedy:

1. Construct utility function g_S by applying the OR construction to g and utility function h_S.
2. Adaptively choose a sequence of items by running Mixed Greedy for utility function g_S with goal value Qm, with respect to the sample distribution $\mathcal{D}_{S,w}$.
3. After goal value Qm is achieved, if the final partial realization b computed by Mixed Greedy does not satisfy $g(b) = Q$, then choose the remaining items in N in a fixed but arbitrary order until $g(b) = Q$.

Theorem 2. *Scenario Mixed Greedy approximates Scenario Submodular Cover with an approximation factor of $O(\log Qm)$, where m is the size of sample S.*

Proof. Scenario Mixed Greedy achieves utility value Q for g when run on any $a \in \Gamma^n$, because the b computed by Mixed Greedy satisfies $a \succeq b$, and the third step ensures Q is reached. Let $c(g)$ and $c(g_S)$ denote the expected cost of the optimal strategies for Scenario SC problems on g and g_S respectively, with respect to sample distribution $\mathcal{D}_{S,w}$. Let τ be an optimal strategy for g achieving expected cost $c(g)$. It is also a valid strategy for the problem on g_S, since it achieves utility Q for g on all realizations, and hence achieves goal utility Qm for g_S on all realizations. Thus $c(g_S) \leq c(g)$.

Functions g and h_S are monotone and submodular. Since g_S is produced from them using the OR construction, g_S is monotone and submodular. Let ρ_S be the value of parameter ρ for the function g_S. By the bound in Theorem 1, running Mixed Greedy on g_S, for the sample distribution $\mathcal{D}_{S,w}$, has expected cost that is at most a $O(\frac{1}{\rho_S} \log Qm)$ factor more than $c(g_S)$. Its expected cost is thus also within an $O(\frac{1}{\rho_S} \log Qm)$ factor of $c(g)$. Making additional choices on realizations not in S, as done in the last step of Scenario Mixed Greedy, does not affect the expected cost, since these realizations have zero probability.

Generalizing an argument from [4], we now prove that ρ_S is lower bounded by a constant fraction. Consider any $b \in (\Gamma \cup \{*\})^n$ and $i \in N$ such that $b_i = *$, and any $\gamma \in \Gamma$ where $\gamma \neq \gamma_{b,i}$. Let $C_b = |S| - h_S(b) = |\{a \in S \mid a \succeq b\}|$. Since sets $\{a \in S \mid a \succeq b$ and $a_i = \gamma\}$ and $\{a \in S \mid a \succeq b$ and $a_i = \gamma_{b,i}\}$ are disjoint, both cannot have size greater than $\frac{C_b}{2}$. Thus $\Delta h_S(b, i, \gamma) \geq \frac{C_b}{2}$ or $\Delta h_S(b, i, \gamma_{b,i}) \geq \frac{C_b}{2}$ or both. By the construction of g_S (recall the definition of the OR construction in (2)), we have that $\Delta g_S(b, i, \gamma) \geq \frac{(Q-g(b))C_b}{2}$ or $\Delta g_S(b, i, \gamma_{b,i}) \geq \frac{(Q-g(b))C_b}{2}$ or both. Since $\gamma_{b,i}$ is the "worst-case" setting for b_i with respect to g_S, it follows that $\Delta g_S(b, i, \gamma) \geq \Delta g_S(b, i, \gamma_{b,i})$, and so in all cases $\Delta g_S(b, i, \gamma) \geq \frac{(Q-g(b))C_b}{2}$. Also, $(Q - g(b))C_b = Qm - g_S(b)$. Therefore, $\rho_S \geq \frac{1}{2}$. The theorem follows from the bound in Theorem 1. □

5   Scenario Adaptive Greedy

Scenario Adaptive Greedy works by first constructing a utility function g_W, produced by applying the OR construction to g and utility function h_W. Here $h_W \colon (\Gamma \cup \{*\})^n \to \mathbb{Z}_{\geq 0}$, where $h_W(b) = W - \sum_{a \in S : a \succeq b} w(a)$. Intuitively, $h_W(b)$ is the total weight of assignments eliminated from S because they are incompatible with the information in b. Utility W is achieved for h_W when all assignments in S have been eliminated. Clearly h_W is monotone and submodular. The function g_W reaches its goal value QW when all possible realizations of the sample have been eliminated or when goal utility is achieved for g. Once g_W is constructed, Scenario Adaptive Greedy runs Adaptive Greedy on g_W.

In an on-line setting, Scenario Adaptive Greedy uses the following procedure to determine the sequence of items to choose on an initially unknown a.

Scenario Adaptive Greedy:

1. Construct modified utility function g_W by applying the OR construction to g and utility function h_W.
2. Run Adaptive Greedy for utility function g_W with goal value QW, with respect to sample distribution $\mathcal{D}_{S,w}$, to determine the choices to make on a.
3. After goal value QW is achieved, if the partial realization b representing the states of the chosen items of a does not satisfy $g(b) = Q$, then choose the remaining items in N in arbitrary order until $g(b) = Q$.

The analysis of Scenario Adaptive Greedy is based on the following lemma.

Lemma 1. *Utility function g_W is adaptive submodular with respect to sample distribution $\mathcal{D}_{S,w}$.*

The proof of Lemma 1 can be found in the full version of the paper.

Theorem 3. *Scenario Adaptive Greedy is an approximation algorithm for Scenario Submodular Cover achieving an approximation factor of $O(\log QW)$, where W is the sum of the weights on the realizations in S.*

Proof. Since g_W is produced by applying the OR construction to g and h_W, which are both monotone, so is g_W. By Lemma 1, g_W is adaptive submodular with respect to the sample distribution. Thus by the bound of Golovin and Krause on Adaptive Greedy, running that algorithm on g_W yields an ordering of choices with expected cost that is at most a $O(\log QW)$ factor more than the optimal expected cost for g_W. By the analogous argument as in the proof of Theorem 2, it follows that Scenario Adaptive Greedy solves the Scenario Submodular Cover problem for g, and achieves an approximation factor of $O(\log QW)$.   □

Acknowledgements. The work in this paper was supported by NSF Grant 1217968. L. Hellerstein thanks Andreas Krause for useful discussions at ETH, and for directing our attention to the bound of Streeter and Golovin for min-sum submodular cover. We thank an anonymous referee for suggesting the Kosaraju trick.

References

1. Bellala, G., Bhavnani, S., Scott, C.: Group-based active query selection for rapid diagnosis in time-critical situations. IEEE Trans. Inf. Theor. **58**, 459–478 (2012)
2. Chen, Y., Javdani, S., Karbasi, A., Bagnell, J.A., Srinivasa, S.S., Krause, A.: Submodular surrogates for value of information. In: Proceedings of the Twenty-Ninth AAAI Conference on Artificial Intelligence, Austin, Texas, USA, 25–30 January 2015, pp. 3511–3518 (2015)
3. Chen, Y., Javdani, S., Karbasi, A., Bagnell, J.A., Srinivasa, S.S., Krause, A.: Submodular surrogates for value of information (long version) (2015). http://las.ethz. ch/files/chen15submsrgtvoi-long.pdf
4. Cicalese, F., Laber, E., Saettler, A.M.: Diagnosis determination: decision trees optimizing simultaneously worst and expected testing cost. In: Proceedings of the 31st International Conference on Machine Learning, pp. 414–422 (2014)
5. Deshpande, A., Hellerstein, L., Kletenik, D.: Approximation algorithms for stochastic boolean function evaluation and stochastic submodular set cover. In: Symposium on Discrete Algorithms (2014)
6. Golovin, D., Krause, A.: Adaptive submodularity: theory and applications in active learning and stochastic optimization. J. Artif. Intell. Res. **42**, 427–486 (2011)
7. Golovin, D., Krause, A., Ray, D.: Near-optimal Bayesian active learning with noisy observations. In: 24th Annual Conference on Neural Information Processing Systems (NIPS), pp. 766–774 (2010)
8. Grammel, N., Hellerstein, L., Kletenik, D., Lin, P.: Scenario submodular cover. CoRR abs/1603.03158 (2016). http://arxiv.org/abs/1603.03158
9. Guillory, A., Bilmes, J.A.: Simultaneous learning and covering with adversarial noise. In: Proceedings of the 28th International Conference on Machine Learning, ICML 2011, Bellevue, Washington, USA, 28 June–2 July 2011, pp. 369–376 (2011)
10. Gupta, A., Nagarajan, V., Ravi, R.: Approximation algorithms for optimal decision trees and adaptive TSP problems. In: Abramsky, S., Gavoille, C., Kirchner, C., Meyer auf der Heide, F., Spirakis, P.G. (eds.) ICALP 2010. LNCS, vol. 6198, pp. 690–701. Springer, Heidelberg (2010). doi:10.1007/978-3-642-14165-2_58
11. Javdani, S., Chen, Y., Karbasi, A., Krause, A., Bagnell, D., Srinivasa, S.S.: Near optimal bayesian active learning for decision making. In: Proceedings of the Seventeenth International Conference on Artificial Intelligence and Statistics, AISTATS 2014, Reykjavik, Iceland, 22–25 April 2014, pp. 430–438 (2014)

12. Kaplan, H., Kushilevitz, E., Mansour, Y.: Learning with attribute costs. In: Symposium on the Theory of Computing, pp. 356–365 (2005)
13. Kosaraju, S.R., Przytycka, T.M., Borgstrom, R.: On an optimal split tree problem. In: Dehne, F., Sack, J.-R., Gupta, A., Tamassia, R. (eds.) WADS 1999. LNCS, vol. 1663, pp. 157–168. Springer, Heidelberg (1999). doi:10.1007/3-540-48447-7_17
14. Navidi, F., Kambadur, P., Nagarajan, V.: Adaptive submodular ranking. CoRR abs/1606.01530 (2016). http://arxiv.org/abs/1606.01530
15. Streeter, M., Golovin, D.: An online algorithm for maximizing submodular functions. In: Advances in Neural Information Processing Systems, pp. 1577–1584 (2009)
16. Ünlüyurt, T.: Sequential testing of complex systems: a review. Discrete Appl. Math. **142**(1–3), 189–205 (2004)
17. Wolsey, L.: Maximising real-valued submodular functions: primal and dual heuristics for location problems. Math. Oper. Res. **7**(3), 410–425 (1982)

Non-greedy Online Steiner Trees on Outerplanar Graphs

Akira Matsubayashi[✉]

Division of Electrical Engineering and Computer Science, Kanazawa University,
Kanazawa 920-1192, Japan
mbayashi@t.kanazawa-u.ac.jp

Abstract. This paper addresses the classical online Steiner tree problem on edge-weighted graphs. It is known that a greedy (nearest neighbor) online algorithm has a tight competitive ratio for wide classes of graphs, such as trees, rings, any class including series-parallel graphs, and unweighted graphs with bounded diameter. However, we did not know any greedy or non-greedy tight deterministic algorithm for other classes of graphs. In this paper, we observe that a greedy algorithm is $\Omega(\log n)$-competitive on outerplanar graphs, where n is the number of vertices, and propose a 5.828-competitive deterministic algorithm on outerplanar graphs. Our algorithm connects a requested vertex and the tree constructed thus far using a path that is constant times longer than the distance between them. The algorithm can be applied to a 21.752-competitive file allocation algorithm against adaptive online adversaries on outerplanar graphs. We also present a lower bound of 4 for arbitrary deterministic online Steiner tree algorithms on outerplanar graphs.

1 Introduction

This paper addresses the classical online Steiner tree problem (STP) on edge-weighted graphs. We are given a graph $G = (V_G, E_G)$ with non-negative edge-weights $w : E_G \to \mathbb{R}^+$ and a subset R of vertices of G. The (offline) Steiner tree problem is to find a *Steiner tree*, i.e., a subtree $T = (V_T, E_T)$ of G that contains all the vertices in R and minimizes its cost $c(T) = \sum_{e \in E_T} w(e)$. In the online version of this problem, vertices $r_1, \ldots, r_{|R|} \in R$ are revealed one by one, and for each $i \geq 1$, we must construct a tree containing r_i by growing the constructed tree for $r_1, \ldots, r_{i-1}$ (null tree for $i = 1$) without information of $r_{i+1}, \ldots, r_{|R|}$.

Imase and Waxman [12] proposed a greedy (nearest neighbor) online algorithm that is $O(\log n)$-competitive on arbitrary graphs with n vertices. They also proved that no deterministic algorithm is $o(\log n)$-competitive even on series-parallel graphs [12]. Westbrook and Yan [15] refined these upper and lower bounds to $\Theta(\log(\mathrm{diam}|R|/\mathrm{OPT}))$ with improving analysis, where diam is the diameter of the underlying graphs, and OPT is the cost of a minimum Steiner tree. The refined upper bound implies that the greedy algorithm is $O(\log \mathrm{diam})$-competitive for unweighted graphs. The greedy algorithm is trivially 1- and 2-competitive on trees and rings, respectively. With these results, the greedy algorithm has a tight competitive ratio for trees, rings, any class of graphs including

© Springer International Publishing AG 2017
K. Jansen and M. Mastrolilli (Eds.): WAOA 2016, LNCS 10138, pp. 129–141, 2017.
DOI: 10.1007/978-3-319-51741-4_11

series-parallel graphs, and unweighted graphs with bounded diameter. However, we did not know any greedy or non-greedy tight deterministic algorithm for other classes of graphs. As for randomized algorithms, a probabilistic embedding of outerplanar graphs into tree metrics with distortion 8, presented by Gupta et al. [11], implies an 8-competitive online Steiner tree algorithm against oblivious adversaries on outerplanar graphs. This embedding was generalized to k-outerplanar graphs with distortion 200^k by Chekuri et al. [9], implying a 200^k-competitive online Steiner tree algorithm against oblivious adversaries on k-outerplanar graphs.

Online Steiner trees on the Euclidean space are studied in [1,4]. Various generalizations of the online Steiner tree problem are also studied, such as generalized STP [5,8,15], asymmetric STP [3,4], priority STP [2,3], and vertex-weighted STP [14].

In this paper, we observe that the greedy algorithm is $\Omega(\log n)$-competitive on outerplanar graphs, and propose a $3+2\sqrt{2} \approx 5.828$-competitive deterministic algorithm on outerplanar graphs. Our algorithm connects a requested vertex and the tree constructed thus far using a path of a maximal length within $\alpha > 1$ times longer than the distance between them. We prove that our algorithm is $(\frac{\alpha}{\alpha-1} + 2\alpha)$-competitive, implying a $(3 + 2\sqrt{2})$-competitive algorithm at $\alpha = 1 + 1/\sqrt{2}$ (Sect. 4). A technical overview will be discussed in the beginning of Sect. 4 after we observe the $\Omega(\log n)$ competitiveness of the greedy algorithm in Sect. 3. Though we do not know if our analysis is tight for $\alpha = 1 + 1/\sqrt{2}$, we observe a lower bound for our algorithm that matches our analysis with any $\alpha \geq 2$. We also present a lower bound of 4 for arbitrary deterministic online Steiner tree algorithms on outerplanar graphs (Sect. 5). Previous and our results are summarized in Table 1.

An application of the online Steiner tree problem is the *file allocation problem*, which is to find dynamic allocations of multiple copies of a data object, called *file*, of size $D \geq 1$ on a network G, such that the total cost of servicing online read/write requests and reallocating the copies is minimized [6,7,13]. Bartal, Fiat, and Rabani [7] proposed a $(2 + \sqrt{3})c$-competitive file allocation algorithm

Table 1. Summary of results of online STP on weighted graphs

Graphs	Competitive ratio	Adversary type	
General graphs	$O(\log n)$	Deterministic	[12]
Series-parallel graphs	$\Omega(\log n)$	Deterministic	[12]
General graphs	$\Theta(\log(\mathrm{diam}|R|/\mathrm{OPT}))$	Deterministic	[15]
Outerplanar graphs	5.828	Deterministic	This paper
Outerplanar graphs	≥ 4	Deterministic	This paper
Rings	2	Deterministic	[7]
Outerplanar graphs	8	Oblivious	[11]
k-outerplanar graphs	200^k	Oblivious	[9]

against adaptive online adversaries based on any c-competitive online Steiner tree algorithm against adaptive online adversaries. Combined with this result, our algorithm implies a $(2 + \sqrt{3})(3 + 2\sqrt{2}) \approx 21.752$-competitive file allocation algorithm against adaptive online adversaries on outerplanar graphs.

2   Preliminaries

Graphs considered here are undirected and have non-negative edge-weights, $w(e) \geq 0$ for any edge e. For a graph G, we denote its vertex set and edge set by V_G and E_G, respectively. We use the notation of w also for graphs, i.e., $w(G) := \sum_{e \in E_G} w(e)$. For a subset R of vertices of G, a *Steiner tree of G for R* is a subtree T of G such that $R \subseteq V_T$. T is said to be *minimum* if T has the minimum cost $w(T)$ overall Steiner trees of G for R.

Suppose that G is a planar graph. The *weak dual* of G is a graph H such that V_H is the set of bounded faces of G, and E_H is the family of sets consisting of two bounded faces that have a common edge. The graph G is *outerplanar* if it can be drawn on the plane so that all the vertices belong to the unbounded face, or equivalently, if H is a forest [10]. We say an edge of G to be *outer* if the edge is contained in the unbounded face, *inner* otherwise.

In the rest of the paper, we assume that G is a biconnected outerplanar graph, because finding a minimum Steiner tree of G can easily be reduced to finding minimum Steiner trees of biconnected components of G. This assumption implies that H is a tree. Let $d_G(u, v)$ be the distance of vertices u and v in G. We use the notation of d_G also for the distance between a graph and a vertex, i.e., $d_G(G', v) := \min\{d_G(u, v) \mid u \in V_{G'}\}$ for a subgraph G' of G and $v \in V_G$.

3   Lower Bound for Greedy Algorithms

In this section, we prove that a greedy algorithm, which always takes a shortest path between a requested vertex and the current tree, is $\Omega(\log n)$-competitive on outerplanar graphs. The proof will be a hint of our algorithm and general lower bound in the following sections. We note that the lower bound is also admitted by another type of greedy algorithm that always takes a shortest path between the current request and one of the previously requested vertices.

Theorem 1. *For any integer $k \geq 0$, there exists a $(2^k + 1)$-vertex outerplanar graph G_k such that if a greedy online Steiner tree algorithm on G_k is ρ-competitive, then $\rho \geq 1 + k/2$.*

Proof. G_k is recursively defined: G_0 consists of two vertices joined by an edge of weight 1. These vertices and edge are said to be of *level* 0. For $i \geq 1$, G_i is obtained from G_{i-1} by adding a new vertex u and edges su and ut of weight $2^{-i} + \epsilon$ for each edge st of level $i - 1$, where $\epsilon > 0$. The added vertices and edges are said to be of *level* i. We illustrate G_4 in Fig. 1.

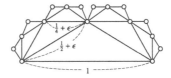

Fig. 1. Outerplanar graph G_4

If all the vertices of G_k are requested in an increasing order with regard to their levels, then a greedy algorithm chooses the unique edge of level 0 first and then 2^{i-1} edges in each level $1 \leq i \leq k$. On the other hand, a minimum Steiner tree has a cost at most that of the tree consisting only of edges of level k. Therefore, the competitive ratio is at least $\frac{1+\sum_{i=1}^{k} 2^{i-1}(2^{-i}+\epsilon)}{2^k(2^{-k}+\epsilon)}$, which tends to $1 + k/2$ as $\epsilon \to 0$. $\square$

4   Algorithm α-Detour and Its Competitiveness

In this section, we define our algorithm, called α-Detour with $\alpha > 1$, and prove its competitiveness.

4.1   Overview

The basic idea of our algorithm is to suppress the cost of a greedy algorithm against the adversary in the proof of Theorem 1. Suppose that the greedy algorithm takes an edge of level i of G_k in the proof of Theorem 1. Then, the vertices incident to the edge can also be connected using edges with level higher than i. For ϵ close to 0, the edge of level i and the detour have nearly the same length. The greedy algorithm incurs an expensive cost of $O(\log n)$ since it takes edges of each level even in such a case. But obviously, we can avoid such an expensive cost by taking the detour with a single penalty of the detour. In our algorithm α-Detour, we take a certain detour of maximal length within the factor of α using edges of higher level. We formally define α-Detour and prove its correctness in Sects. 4.2 and 4.3, respectively. We will introduce in the definition a rooted forest structure of edges with regard to their levels. In G_k, for example, edges su and ut of level i added to an edge st of level $i - 1$ are children of st.

In our analysis of the competitiveness, intuitively (not precisely), we charge the weight of each edge uv chosen by α-Detour to the path connecting u and v in a minimum Steiner tree. For an edge e of the minimum Steiner tree, edges charged to e are of three types: (i) ancestor edges of e, (ii) descendant edges of e or e itself, and (iii) otherwise. The amounts charged to e by edges of type (i) are essentially fragments of weights exponentially decreasing by the factor α^{-1}. Hence, the total charged amount of this type is at most $\sum_{i \geq 1} \alpha^{-(i-1)} w(e) < \frac{\alpha}{\alpha-1} w(e)$. For each of types (ii) and (iii), by the property of detours based on shortest paths, the total charged amount is at most $w(e)$ multiplied by α. Summing these amounts, we derive that α-Detour is $(\frac{\alpha}{\alpha-1}+2\alpha)$-competitive as desired. We formally prove this in Sect. 4.4. One non-intuitive part of the proof is that we charge an edge uv chosen by α-Detour to not the whole but only a part of the path joining uv in the minimum Steiner tree. Specifically, we do not charge uv to any part of the tree that was constructed by α-Detour before uv is chosen. The charging process is defined through dynamically modifying the graph and forest structure of edges, in such a way that the forest precisely represents the relation between shortest paths and their detours.

4.2   Definition

For the first requested vertex r_1, we suppose that the weak dual H of G is a tree rooted by a face containing r_1. We introduce a forest F with $V_F = E_G$ as follows. If a face C is the root of H, then all the edges of C are the roots of the connected components of F. If C is a face of G, and C' is a child of C in H, then all the edges in $E_{C'} \setminus E_C$ are the children of the unique edge $e \in E_C \cap E_{C'}$ in F. For any inner edge e of G, let G_e be the subgraph of G induced by the descendant edges of e in F. We note that G_e does not contain e. To clarify our discussion, we use the term *links* to denote elements of E_F.

For the ith requested vertex r_i, α-Detour constructs a tree T_i as follows:

α-Detour

1. If $i = 1$, then return the tree T_1 consisting only of r_1.
2. Suppose $i \geq 2$. If $r_i \in V_{T_{i-1}}$, then return $T_i := T_{i-1}$.
3. Otherwise, find a shortest path $P_i = (p_1, p_2, \ldots, p_{|P_i|})$ between a vertex p_1 in T_{i-1} and $p_{|P_i|} = r_i$.
4. For $j = 1$ to $|P_i|-1$, if $p_{j+1} \notin V_{T_{i-1}}$, then call DetourEdge$(\alpha, p_j, p_{j+1})$ defined below.
5. Return $T_i := T_{i-1}$.

DetourEdge(β, u, v) is a procedure to modify T_{i-1} by adding a path between T_{i-1} and v of maximal length at most $\beta \cdot w(uv)$. The inputs are $\beta \geq 1$ and an edge uv such that $u \in V_{T_{i-1}}$, $v \notin V_{T_{i-1}}$, and $w(uv) \leq d_G(T_{i-1}, v)$. The procedure is formally defined as follows:

DetourEdge(β, u, v)

1. If uv is outer, then add uv to T_{i-1} and return.
2. If uv is inner, then find a shortest path $Q_{uv} = (q_1, \ldots, q_{|Q_{uv}|})$ from a vertex q_1 in T_{i-1} to $q_{|Q_{uv}|} = v$ in G_{uv}.
3. If $w(Q_{uv})/w(uv) > \beta$, then add uv to T_{i-1}.
4. Otherwise, call DetourEdge$(\beta \cdot w(uv)/w(Q_{uv}), q_j, q_{j+1})$ for $j = 1$ to $|Q_{uv}|-1$.
5. Return.

4.3   Correctness

Since α-Detour and DetourEdge only add edges to T_{i-1}, T_i contains T_{i-1} as a subgraph. Therefore, it suffices to show that α-Detour connects r_i to T_{i-1} with a path of length at most $\alpha \cdot d_G(T_{i-1}, r_i)$.

Lemma 1. *DetourEdge*(β, u, v) *adds a path of length at most* $\beta \cdot w(uv)$ *between a vertex of* T_{i-1} *and* v.

Proof Sketch. Induction on the height of uv in F.   □

Since α-Detour calls DetourEdge(α, p_j, p_{j+1}) for each j unless p_{j+1} has already been contained in T_{i-1}, by Lemma 1, we have the following lemma:

Lemma 2. *For* $i \geq 2$, α-*Detour connects* r_i *to* T_{i-1} *with a path of length at most* $\alpha \cdot d_G(T_{i-1}, r_i)$.

4.4   Competitiveness

We first introduce dynamic modification of G and F in such a way that any Q_{uv} found in DetourEdge(β, u, v) is a path connecting u and v. According to the modified graph and forest, we then charge the weights of edges of P_i in Step 3 of α-Detour to their descendant edges that are potentially used by a minimum Steiner Tree. We finally estimate the competitiveness by comparing the charged amounts multiplied by α, including the case that edges of P_i are charged not to their descendants, to the cost of the minimum Steiner Tree.

Modifying Graph. Every time DetourEdge(α, p_j, p_{j+1}) is called in Step 4 of α-Detour, we mark $p_j p_{j+1}$ "greedy". Before entering DetourEdge(α, p_j, p_{j+1}), we perform the following:

ModifyGraph

1. For each "greedy" edge pp' such that $p_j p_{j+1}$ is an ancestor of pp', and that there is no "greedy" edge that is a descendant of $p_j p_{j+1}$ and an ancestor of pp', we decompose the graph into the subgraph induced by pp' and its all descendants and the subgraph induced by other edges. We then contract the latter subgraph by identifying p and p'. If there is a self-loop of the identified vertex, then we remove it.
2. According to the modified graph, we modify the forest structure as well. I.e., we remove the link of the forest between pp' and its parent. This yields a subtree rooted by pp'. We then remove from the forest any self-loop of the identified vertex.

We note that if we performed ModifyGraph before calling DetourEdge (α, p_j, p_{j+1}) in Step 4 of α-Detour, then DetourEdge(α, p_j, p_{j+1}) would choose the same edges of the path between the current Steiner tree and p_{j+1}. Moreover, the path P_i chosen in Step 3 of α-Detour is not affected either. These are because we decompose or contract the graph only at "greedy" edges whose end-vertices are already contained in the Steiner tree. Therefore, we may discuss as if α-Detour and DetourEdge were performed with modifying the graph. We observe some properties related to the modification of the graph as stated in the following lemmas. To clarify our discussion, we use G and F to denote the initial graph and forest before processing r_1, respectively, and G^* and F^* to denote the final graph and forest after processing $r_1, \ldots, r_{|R|}$, respectively. We also use G_e^* just as defined for G.

Lemma 3. *For any edge uv such that $DetourEdge(\beta, u, v)$ is called, Q_{uv} is a path connecting u and v in the modified graph at the point that $DetourEdge(\beta, u, v)$ is processed.*

Proof Sketch. Previous ModifyGraph for some "greedy" edge makes the current Steiner tree into a single vertex in the subgraph induced by descendant edges of the "greedy" edge. After that, any DetourEdge for an edge e constructs a path using only descendant edges of e. Since uv is not a descendant of such e, this means that when $DetourEdge(\beta, u, v)$ is processed, u is the unique vertex of the current Steiner tree in the subgraph induced by the descendants of uv. □

Lemma 4. *For any edge uv such that $DetourEdge(\beta, u, v)$ is called, uv and edges of Q_{uv} are contained in the same connected component of G^*.*

Proof Sketch. By Lemma 3, Q_{uv} connects u and v at the point that $DetourEdge(\beta, u, v)$ is processed. After that, therefore, any edge that is a descendant of uv and an ancestor of an edge of Q_{uv} cannot be "greedy". □

Lemma 5. *For any edge uv such that $DetourEdge(\beta, u, v)$ is called, Q_{uv} is a shortest path between u and v in G_{uv}^*.*

Proof Sketch. By Lemma 4, uv and Q_{uv} are contained in the same connected component of G^*. If there is a path Q' between u and v in G_{uv}^* shorter than Q_{uv}, then ModifyGraph for some "greedy" descendant pp' of uv must shorten Q' so that $w(Q') < w(Q_{uv})$. This means that before this ModifyGraph, Q' contained a subpath between p and p' with descendant edges of pp', some of which are removed by the ModifyGraph. However, since pp' is "greedy", the resulting subpath cannot be shorter than pp'. This means that ModifyGraph does not make Q' shorter than Q_{uv}. □

By Lemma 5, we immediately have the following:

Lemma 6. *For any edge e' in Q_e for some edge e, $w(e')$ is at most the distance between the end-vertices of e' in $G_{e'}^*$.*

Charging Weights. For a "greedy" edge e, we define the amount charged to any descendant e' of e in F^* as $w(e)$ multiplied by a factor $f_{e \to e'}$. Essentially, $f_{e \to e'}$ is defined as the ratio $w(e')/w(Q_e)$ for e' in Q_e. Moreover, we define the factor to be transitive, i.e., $f_{e \to e'} = f_{e \to e''} \cdot f_{e'' \to e'}$ if $f_{e \to e''}$ and $f_{e'' \to e'}$ are defined. To extend this definition to any e', we extend the notion of Q_e to any edge e.

We formally define the factor as follows: For any edge e in G^*, let S_e be a shortest path connecting the end-vertices of e in G_e^* if e is inner, e itself otherwise. We note that if e has Q_e, then $w(Q_e) = w(S_e)$ by Lemma 5. Suppose that e and e' are an edge and its descendant in F^*, respectively. If e' is in S_e, then let $f_{e \to e'} := w(e')/w(S_e)$. If e' is an ancestor of an edge of S_e, then let $f_{e \to e'} := w(S_{e'})/w(S_e)$. We note that $S_{e'}$ is a subpath of S_e in this case. If e' is a descendant of an edge of S_e, then there is a sequence of edges $e_1 = e, e_2, \ldots, e_h = e'$ such that e_{i+1} is in S_{e_i} for $1 \le i \le h - 2$, and either e_h is in $S_{e_{h-1}}$ or S_{e_h} is a subpath of $S_{e_{h-1}}$. For such e', we define $f_{e \to e'} := \prod_{i=1}^{h-1} f_{e_i \to e_{i+1}}$. In addition, we define $f_{e \to e} := 1$.

Lemma 7. *For any edge e and a path P connecting the end-vertices of e in G_e^*, it follows that $\sum_{e' \in E_P} f_{e \to e'} = 1$.*

Proof Sketch. Induction on the height of e in F^*. □

Lemma 8. *For any edge e and its descendant e' in F^*, $f_{e \to e'} \le w(e')/w(S_e)$.*

Proof Sketch. By the definition of $f_{e \to e'}$ and Lemma 6. □

Lemma 9. *Suppose that uv is a "greedy" edge, and that D_{uv} is the path between the Steiner tree and v constructed by DetourEdge(α, u, v) in Step 4 of α-Detour. If e is an edge in $G_{e'}^*$ for an edge e' in D_{uv}, then $w(e) > \alpha f_{uv \to e} w(uv)$.*

Proof Sketch. We prove the lemma by induction on the number of recursive levels for DetourEdge(α, u, v) to output e'. For the base case, i.e., $uv = e'$, by Lemmas 8 and 5 and $w(Q_{uv})/w(uv) > \alpha$, we can obtain $w(e) > \alpha f_{uv \to e} w(uv)$. For an induction step, assuming DetourEdge(β, u', v') called with $\beta = \alpha \cdot w(uv)/w(Q_{uv})$ for some edge $u'v'$ in Q_{uv}, we can obtain $w(e) > \alpha f_{uv \to e} w(uv)$ by Lemma 5. □

Comparison to Minimum Steiner Tree. Suppose that Z is any Steiner tree for R in G. Our aim is to decompose G into subgraphs according to Z, associate "greedy" edges with the decomposed subgraphs, and to estimate the amount charged to the edges of Z by "greedy" edges in each decomposed subgraph.

Specifically, for any edge e of Z, we decompose G into the subgraph induced by e and its descendant in F and the subgraph induced by edges that are not descendants of e in F. Decomposing G by all edges of Z, we obtain a set $\mathcal{B}$ of outerplanar subgraphs of G, each of which has edges of Z only in its unbounded face. For a subgraph $B \in \mathcal{B}$, B has either at most one edge or the all edges of the root face of G. If B has at most one edge of the root face, then B has a root edge e_B in B, i.e., an ancestor of all the other edges of B in F. We note that Z has e_B. If B has the entire root face, then for convenience, we suppose that Z has a null edge $e_B = r_1 r_1$ with the weight of 0, and that e_B is the parent of the other edges in the root face. I.e., we suppose that e_B is an ancestor of all the other edges of B also in this case. Let Z_B be the path induced by $E_B \cap E_Z$.

We associate a "greedy" edge uv with B if $E_B \setminus E_Z \cup \{e_B\}$ contains an edge of D_{uv}, where D_{uv} is the path between the Steiner tree and v constructed by DetourEdge(α, u, v). A "greedy" edge is said to be *open in B* if the edge is associated with B, and is an outer edge in $E_B \setminus E_Z$ or an ancestor of an outer edge in $E_B \setminus E_Z$ in F^*. A "greedy" edge associated with B and not open in B is said to be *closed in B*. In other words, all the outer edges of B that are descendants in F^* of a "greedy" edge closed in B are contained in Z.

Lemma 10. *For any edge z in $Z_B \setminus e_B$, the total amount charged to z by "greedy" edges associated with B is less than $w(z)/(\alpha - 1)$.*

Proof Sketch. Let $e_1, \ldots, e_h$ be the "greedy" edges associated with B such that in F^*, e_i is an ancestor of e_{i+1} for $1 \le i < h$, and e_h is an ancestor of z. By Lemma 9, we can prove that the amount charged to z by e_i is less than $\frac{w(z)}{\alpha^{h-i+1}}$. Summing this overall i, we have the lemma.   □

Lemma 11. *Let X_B be the set of "greedy" edges open in B. If there is $x \in X_B \setminus E_B$, then $w(X_B \setminus x) + f_{x \to e_B} w(x) \le w(Z_B)$. Otherwise, $w(X_B) \le w(Z_B)$.*

Proof. Let $X_B = \{e_1, \ldots, e_{|X_B|}\}$, and suppose that for each $i \ge 1$, e_i is marked "greedy" earlier than e_{i+1} is. We first assume that $X_B \setminus E_B = \emptyset$. When e_1 is marked "greedy", e_1 is contained in a shortest path P from the current Steiner tree to a request vertex. Since e_1 is the first open edge marked "greedy", e_1 is incident to at least one vertex s_1 of Z_B. For otherwise, the current Steiner tree has a vertex not incident to an edge of Z_B, which implies that there must be an edge open in B and marked "greedy" earlier than e_1. Because P must reach the request vertex, which is contained in Z_B, we can find the vertex $t_1 \in V_{Z_B} \cap V_P \setminus s_1$ nearest to s_1 on P. We note that the subpath P^1 of P between s_1 and t_1 consists only of open edges, say $e_1, \ldots, e_j$. This is justified through observing that e_h ($1 < h \le j$) cannot be closed if e_{h-1} is open and the vertex incident to both e_h and e_{h-1} is not in Z_B. We charge the weights of $e_1, \ldots, e_j$ to the subpath Z_B^1 of Z_B between s_1 and t_1. We note that since P is a shortest path, the total charged amount is at most $w(Z_B^1)$.

We continue the similar process for the remaining edges of X_B. I.e., we find the vertex t_2 in Z_B nearest to e_{j+1} on a shortest path P' from the current Steiner tree to a request vertex. One exception is that since e_{j+1} is not the first open edge in X_B, e_{j+1} may join two vertices in $V_B \setminus V_{Z_B}$. In this case, we set s_2 to the vertex in $\{s_1, t_1\}$ that is closer to t_2. If e_{j+1} is incident to a vertex in Z_B, then we set s_2 to the vertex as done for s_1. We charge the weights of the open edges in the subpath P^2 of P' from e_{j+1} to t_2 to the subpath Z_B^2 of Z_B between s_2 and t_2. We note again that P^2 consists only of open edges, and that the total charged amount is at most $w(Z_B^2)$. Moreover, Z_B^1 and Z_B^2 are edge-disjoint. For otherwise, s_2 or t_2 is in $V_{Z_B^1} \setminus \{s_1, t_1\}$. In either case, e_{j+1} joins two vertices of the cycle formed by P^1 and Z_B^1. By the definition of ModifyGraph, all the edges of P^1 that are descendants of e_{j+1} in F are identified to a single vertex in the subgraph of G^* containing e_{j+1}. These situations imply that e_{j+1} is neither an outer edge in $E_B \setminus E_{Z}$ nor an ancestor of such an outer edge in F^*, contradicting that e_{j+1} is open in B. Repeating this process, we can charge all the edges in X_B to Z_B in such a way that $w(X_B) \le w(Z_B)$.

We then assume that there is $x \in X_B \setminus E_B$. Such x is an ancestor of e_B and unique. We process $e_1, e_2, \ldots$ as describe above, except that we charge $f_{x \to e_B} w(x)$ to e_B for x, i.e., e_B is the subpath Z_B^i for some i when processing x. We observe that the subpaths $Z_B^1, \ldots, Z_B^{i-1}$ do not contain e_B. For otherwise, there is a subpath $P^{i'}$ with $i' < i$ such that $Z_B^{i'}$ contains e_B. When x is marked "greedy", $P^{i'}$ is identified to a single vertex in the subgraph of G^* containing x. This implies that x is not open in B. We also observe that the subpaths $Z_B^{i+1}, \ldots$ do not contain e_B. This is because the assumption that x is associated with B implies that the end-vertices of e_B are added to α-Detour's Steiner tree when x is detoured. Thus, we can charge all the edges in X_B to Z_B in such a way that $w(X_B \setminus x) + f_{x \to e_B} w(x) \leq w(Z_B)$.   □

Lemma 12. *It follows that $w(T_{|R|}) < (\alpha/(\alpha - 1) + 2\alpha) w(Z)$.*

Proof Sketch. We can observe that each "greedy" edge e is associated with at least one subgraph B in $\mathcal{B}$. Moreover, if e is open in B, then $w(e)$ is directly charged to Z_B as described in Lemma 11. If e is closed in B, then $w(e)$ is fully charged to Z_B by Lemma 7. Thus, if we denote the set of "greedy" edges closed in B by Y_B, then it follows from Lemma 2 that

$$w(T_{|R|}) \leq \alpha \sum_{B \in \mathcal{B}} \left[\sum_{e \in Y_B} w(e) \right.$$
$$\left. + \begin{cases} \sum_{e \in X_B \setminus x} w(e) + f_{x \to e_B} w(x) & \text{if } \exists x \in X_B \setminus E_B \\ \sum_{e \in X_B} w(e) & \text{otherwise} \end{cases} \right].$$

Applying Lemmas 10 and 11,

$$w(T_{|R|}) \leq \alpha \sum_{B \in \mathcal{B}} \left[\sum_{z \in E_{Z_B} \setminus e_B} \frac{w(z)}{\alpha - 1} + w(Z_B) \right]$$
$$= \alpha \sum_{B \in \mathcal{B}} \left[\frac{w(Z_B \setminus e_B)}{\alpha - 1} + w(Z_B \setminus e_B) + w(e_B) \right]$$
$$= \alpha \sum_{B \in \mathcal{B}} \left[\left(\frac{1}{\alpha - 1} + 1 \right) w(Z_B \setminus e_B) + w(e_B) \right]$$
$$= \left(\frac{\alpha}{\alpha - 1} + 2\alpha \right) w(Z).$$

□

Setting $\alpha = 1 + 1/\sqrt{2}$, we have the following theorem.

Theorem 2. *Algorithm $(1 + 1/\sqrt{2})$-Detour is $3 + 2\sqrt{2} \approx 5.828$-competitive.*

We do not know if the upper bound of 5.828 is tight. However, our analysis of Lemma 12 is tight for $\alpha \geq 2$.

Theorem 3. *For any $\alpha > 1$, there exists an outerplanar graph G_α such that if α-Detour is ρ-competitive on G_α, then $\rho \geq \min\{3\alpha, \alpha/(\alpha - 1) + 2\alpha\}$.*

5   Lower Bound for Arbitrarily Algorithms

In this section, we prove a lower bound of 4 for any deterministic Steiner tree algorithm on outerplanar graphs.

Overview. Our idea is based on Theorem 1. We recursively define a class of outerplanar graphs just as done in the proof of the theorem, except that arbitrarily many vertices and edges of level i are added to an edge of level $i - 1$. Our adversary generates requests in phases; in the ith phase, vertices along the children of the online algorithm's tree up to the $(i - 1)$st phase are requested. An online algorithm possibly chooses detours with (variable) factor α. The key of our proof is to define a sequence of upper bounds γ_i of α to be ρ-competitive at the ith phase. In fact, we can prove that for any $\rho < 4$, there is i such that $\gamma_i = 1$. This means that if there is a ρ-competitive algorithm with $\rho < 4$, then it tends to a greedy algorithm; however, this is impossible.

Definition of Graph. Let m be a positive integer and ϵ be a positive real number. Let G_0 be a path consisting of a single edge of weight 1. The unique edge of G_0 is said to be of level 0. For $i \geq 1$, let G_i be the graph obtained from G_{i-1} by adding m^i edges of weight $(1 + \epsilon)^i / \prod_{j=1}^{i} m^j$ to each edge of level $i - 1$ in such a way that the added m^i edges form a path connecting the end-vertices of the edge of level $i-1$. All the added edges are said to be of level i. We suppose $G := G_i$ with sufficiently large i. We define F as the rooted tree with $V_F = E_G$ such that for an edge e of level $i - 1$, m^i edges added to e are children of e in F. We note that such children has the total weight of $(1 + \epsilon)w(e)$.

Adversary. We use a sequence K_i for $i \geq 0$ defined as follows: Let $K_0 := 1$ and K_1 be less than but sufficiently close to 3. For $i \geq 1$, we define $K_{i+1} := (K_0 + K_1)(K_i - K_{i-1})$ if $K_i < (K_0 + K_1)(K_i - K_{i-1})$, and $K_{i+1} := K_i$ if $K_i \geq (K_0 + K_1)(K_i - K_{i-1})$.

Our adversary ADV generates a request sequence against a deterministic Steiner tree algorithm ALG on G. In the initial phase, called the 0th phase, ADV defines $Z_0 := G_0$ and requests vertices of Z_0. Let T_0 be the Steiner tree computed by ALG for these requests, and P_0 be the path in T_0 connecting the requests. For the ith phase with $i \geq 1$, ADV defines the path Z_i consisting of children of edges of P_{i-1}, and requests vertices of Z_i that have not been requested. Let T_i be the Steiner tree computed by ALG for all the requested vertices thus far. For an edge e in P_{i-1}, vertices incident to a child of e must be contained in the subgraph S of T_i induced by the descendants of e. If S is connected, then there is a path Q_e in S connecting the end-vertices of e. Otherwise, since T_i is connected, there is a unique child m_e such that $S \cup m_e$ has a path Q_e connecting the end-vertices of e. Let P_i be the path obtained by concatenating Q_e for all edges e in P_{i-1}. We can inductively observe that P_i and Z_i are Steiner trees for the requests up to the ith phase. If $w(P_i) > \gamma_i w(P_{i-1})$, then ADV quits generating requests, where $\gamma_i := K_i/K_{i-1} \geq 1$. Otherwise, ALG performs the next phase.

Analysis. The following lemma is used to guarantee that ADV quits in finite phases.

Lemma 13. *There exists $\ell \geq 1$ such that $K_{\ell+1} = K_\ell$.*

Proof Sketch. Observing that a sequence $(a_i)_{i \geq 0}$ with the recurrence $a_{i+1} = b(a_i - a_{i-1})$ oscillates for $0 < b < 4$, we have the lemma. $\square$

Lemma 13 implies $\gamma_{\ell+1} = K_{\ell+1}/K_\ell = 1$. On the other hand,

$$w(P_i) \geq w(Z_i) = (1+\epsilon)w(P_{i-1}) \tag{1}$$

by the definitions of P_i and Z_i. Therefore, ADV performs at most $\ell+1$ phases.

The following lemma is used to estimate the ratio of the cost of ALG to the cost of ADV.

Lemma 14. $\sum_{i=0}^{j} K_i/K_{j-1} \geq K_0 + K_1$ *for any $j \geq 1$.*

Proof Sketch. Induction on j. $\square$

Lemma 15. *If ADV quits at the qth phase, then $w(T_q)/w(Z_q)$ tends to 4 as $m \to \infty$, $\epsilon \to 0$, and $K_1 \to 3$.*

Proof. By definition, P_i consists of descendants of edges in P_{i-1}. This means that P_i and P_{i-1} are edge-disjoint. Therefore, it follows that $w(T_j) \geq \sum_{i=0}^{q} w(P_i) - \delta$, where δ is the sum of $w(m_e)$ overall edges e in $P_0, \ldots, P_{q-1}$ having m_e. We can upper bound δ by summing weight of one child of all edges; therefore,

$$\delta \leq \sum_{i \geq 1} \left(\prod_{j=1}^{i-1} m^j \right) \frac{(1+\epsilon)^i}{\prod_{j=1}^{i} m^j} = \sum_{i \geq 1} \left(\frac{1+\epsilon}{m} \right)^i < \frac{\frac{1+\epsilon}{m}}{1 - \frac{1+\epsilon}{m}} \to 0 \quad [m \to \infty].$$

Since ADV quits at the qth phase, it follows that $w(P_i) \leq \gamma_i w(P_{i-1})$ for $1 \leq i < q$ and $w(P_q) > \gamma_q w(P_{q-1})$. Therefore, it follows from Lemma 14 that

$$\lim_{m \to \infty} \frac{w(T_q)}{w(Z_q)} = \frac{\sum_{i=0}^{q} w(P_i)}{w(Z_q)} \geq \frac{\sum_{i=0}^{q-1} w(P_i) + w(P_q)}{(1+\epsilon)w(P_{q-1})} \quad [\text{by (1)}]$$

$$> \frac{\sum_{i=0}^{q-1} \prod_{j=i}^{q-2} \gamma_{j+1}^{-1} w(P_{q-1})}{(1+\epsilon)w(P_{q-1})} + \frac{\gamma_{q-1}}{1+\epsilon} = \frac{1}{1+\epsilon} \left(\frac{\sum_{i=0}^{q-1} K_i}{K_{q-1}} + \frac{K_q}{K_{q-1}} \right)$$

$$\geq \frac{K_0 + K_1}{1+\epsilon} \to 4. \quad [\epsilon \to 0, K_1 \to 3, K_0 = 1]$$

$\square$

Thus, we have the following theorem.

Theorem 4. *If a deterministic online Steiner tree algorithm is ρ-competitive on outerplanar graphs, then $\rho \geq 4$.*

References

1. Alon, N., Azar, Y.: On-line steiner trees in the Euclidean plane. Discret. Comput. Geom. **10**, 113–121 (1993)
2. Angelopoulos, S.: Online priority steiner tree problems. In: Dehne, F., Gavrilova, M., Sack, J.-R., Tóth, C.D. (eds.) WADS 2009. LNCS, vol. 5664, pp. 37–48. Springer, Heidelberg (2009). doi:10.1007/978-3-642-03367-4_4
3. Angelopoulos, S.: Parameterized analysis of online steiner tree problems. In: Adaptive, Output Sensitive, Online and Parameterized Algorithms. Dagstuhl Seminar Proceedings (2009). http://drops.dagstuhl.de/opus/volltexte/2009/2121
4. Angelopoulos, S.: On the competitiveness of the online asymmetric and Euclidean steiner tree problems. In: Bampis, E., Jansen, K. (eds.) WAOA 2009. LNCS, vol. 5893, pp. 1–12. Springer, Heidelberg (2010). doi:10.1007/978-3-642-12450-1_1
5. Averbuch, B., Azar, Y., Bartal, Y.: On-line generalized steiner problem. Theor. Comput. Sci. **324**, 313–324 (2004)
6. Awerbuch, B., Bartal, Y., Fiat, A.: Competitive distributed file allocation. Inf. Comput. **185**(1), 1–40 (2003)
7. Bartal, Y., Fiat, A., Rabani, Y.: Competitive algorithms for distributed data management. J. Comput. Syst. Sci. **51**(3), 341–358 (1995)
8. Berman, P., Coulston, C.: On-line algorithms for steiner tree problems. In: Proceedings of the 29th ACM Symposium on Theory of Computing, pp. 344–353 (1997)
9. Chekuri, C., Gupta, A., Newman, I., Rabinovich, Y., Sinclair, A.: Embedding k-outerplanar graphs into ℓ_1. SIAM J. Discret. Math. **20**(1), 119–136 (2006)
10. Fleischner, H.J., Geller, D.P., Harary, F.: Outerplanar graphs and weak duals. J. Indian Math. Soc. **38**, 215–219 (1974)
11. Gupta, A., Newman, I., Rabinovich, Y., Sinclair, A.: Cuts, trees, and ℓ_1-embedding of graphs. Combinatorica **24**(2), 233–269 (2004)
12. Imase, M., Waxman, B.M.: Dynamic steiner tree problem. SIAM J. Discret. Math. **4**(3), 369–384 (1991)
13. Lund, C., Reingold, N., Westbrook, J., Yan, D.: Competitive on-line algorithms for distributed data management. SIAM J. Comput. **28**(3), 1086–1111 (1999)
14. Naor, J.S., Panigrahi, D., Singh, M.: Online node-weighted steiner tree and related problems. In: Proceedings 52nd Annual IEEE Symposium on Foundations of Computer Science, pp. 210–219 (2011)
15. Westbrook, J., Yan, D.C.K.: The performance of greedy algorithms for the on-line steiner tree and related problems. Math. Syst. Theory **28**, 451–468 (1995)

A Refined Analysis of Online Path Coloring in Trees

Astha Chauhan and N.S. Narayanaswamy$^{(\boxtimes)}$

Department of Computer Science and Engineering,
Indian Institute of Technology Madras, Chennai, India
{astha,swamy}@cse.iitm.ac.in

Abstract. Our results are on the online version of path coloring in trees where each request is a path to be colored online, and two paths that share an edge must get different colors. For each T, we come up with a hierarchical partitioning of its edges with a minimum number of parts, denoted by $h(T)$, and design an $O(h(T))$-competitive online algorithm. We then use the lower bound technique of Bartal and Leonardi [1] along with a structural property of the hierarchical partitioning, to show a lower bound of $\Omega(h(T)/\log(4h(T)))$ for each tree T on the competitive ratio of any deterministic online algorithm for the problem. This gives us an insight into online coloring of paths on *each* tree T, whereas the current tight lower bound results are known only for special trees like paths and complete binary trees.

1 Introduction

The problem of path coloring in graphs has been motivated by the problem of wavelength allocation in communication networks that make use of Wavelength Division Multiplexing (WDM). In WDM, multiple optical signals are transmitted simultaneously through the same fibre link but at different wavelengths of light. Any two nodes in such a network communicate by establishing a path between them and assigning a wavelength to the path. Paths which use the same fibre link are assigned different wavelengths. A natural goal is to minimize the number of wavelengths used in such a network. This crucial problem in communication networks is known as the *wavelength allocation problem*: Given a network and a set of requests on the network, the problem is to assign distinct wavelengths to all requests that share a communication link. This problem may be viewed as the problem of coloring paths on a network graph such that two paths that share a link receive different colors (representing wavelengths in communication network). One of the most well-studied network topologies in this framework is the tree topology. Now, we formally define the *path coloring problem on trees*:

© Springer International Publishing AG 2017
K. Jansen and M. Mastrolilli (Eds.): WAOA 2016, LNCS 10138, pp. 142–154, 2017.
DOI: 10.1007/978-3-319-51741-4_12

PATH COLORING ON TREES

Instance: A tree T and a set $\mathcal{P}$ of paths of T

Output: A coloring function $c : \mathcal{P} \mapsto \{1, \cdots, r\}$ for some integer $r \geq 1$ such that for every pair of distinct paths P_i and P_j in $\mathcal{P}$ with $P_i \cap P_j \neq \emptyset$, $c(P_i) \neq c(P_j)$

Goal: To obtain a function c which minimizes r.

Several researchers have extensively studied both offline and online versions of this problem on different network topologies. In the offline setting, the topology, and the entire request sequence are known in advance. However, in online wavelength allocation problems, though the topology is known in advance, requests arrive one at a time and a request has to be assigned a wavelength as soon as it is presented (this assignment cannot be changed later). In an online algorithm, the inputs arrive in a sequence and each input has to be processed depending only on already received requests and served as soon as it arrives with no knowledge of future requests. The performance of an online algorithm A is analyzed using the *competitive ratio*. It is the worst-case ratio between the cost of the solution found by the algorithm A to the cost of an optimal solution.

The path coloring problem for trees may be viewed as the vertex coloring problem for edge intersection graph of paths on a tree, called *EPT graphs*, such that no two adjacent vertices in the intersection graph get the same color. In the edge intersection graph of the given paths, two vertices are adjacent if and only if the corresponding paths have a common edge. The vertex intersection graphs of paths of a tree is the class of path graphs [5], which is a subclass of chordal graphs. Chordal graphs are the vertex intersection graphs of subtrees of a tree [4] and can be optimally colored in polynomial time [3]. Thus, in the offline setting, path graphs can be optimally colored in polynomial time. However, coloring of edge intersection graph of paths in an undirected tree (EPT graph) has been shown to be NP-complete [7]. Tarjan [16] gave a $\frac{3}{2}$-approximation algorithm for coloring EPT graphs. Erlebach and Jansen [2] showed that in the case of undirected trees of bounded degree, the path coloring problem can be solved in polynomial time. However, for undirected trees of arbitrary degree, the problem is NP-hard and approximation algorithm with absolute approximation ratio $\frac{4}{3}$ and asymptotic approximation ratio $\frac{11}{10}$ are known [2]. Path coloring is also proved to be NP-hard on undirected and bi-directed ring networks [2], bi-directed binary trees [2] and bi-directed binary caterpillars [15]. Several interesting approaches have been proposed in the literature([2,8,10,11,14,15]) for network topologies like rings, caterpillars, trees and trees of rings. If the tree itself is a path, then the edge intersection graph of subpaths of this path is an interval graph [6]. Therefore, in this case, coloring algorithms for the vertex intersection graphs can be used for optimally coloring the edge intersection graphs. The coloring of the vertex intersection graphs on the line topology (special case of tree topology, i.e., a tree in which no vertex has degree 3 or more) is very well studied. It is known to be optimally polynomial time solvable in the offline setting.

The path coloring problem is also extensively studied in the online framework. Bartal and Leonardi [1] proposed an $O(\log(n))$ competitive online algorithm (described in Sect. 2 and subsequently called as the BL algorithm) for coloring edge intersection graphs of paths on a tree. They also [1] gave a lower bound of $\Omega(\frac{\log n}{\log \log n})$ for the online path coloring problem on complete binary trees. The path coloring problem on the line topology has been studied in the context of interval graph coloring. The edge intersection graph and vertex intersection graph of paths on line topology are both known to be interval graphs [6]. Thus, the problem of online coloring paths on the line topology is equivalent to the problem of online coloring vertices of interval graph which is a subclass of chordal graphs. One of the simplest strategies adopted for the online coloring of vertices in an interval graph is the First Fit strategy: allocate the color of least index permissible. Several researchers have analyzed this method and have come up with improved bounds over the years and 8 is the best known competitive ratio [12]. However, a 3-competitive recursive algorithm by Kierstead and Trotter [9] (subsequently called as the KT algorithm) for the online coloring of interval graphs, was known much earlier (a detailed presentation of the algorithm can be found in [13]). They showed a matching lower bound as well; that is, no deterministic online algorithm can achieve a competitive ratio better than 3.

1.1   Our Motivation and Results

We address the problem of online path coloring in trees and design algorithm that uses the Kierstead-Trotter approach [9] to color paths in the line topology. Throughout the paper, we consider the coloring problem on edge intersection graphs. For the online path coloring in the line topology, we use the optimal KT algorithm on the edge intersection graph, which is an interval graph. The starting point of our work is the BL Algorithm [1], which achieves a competitive ratio of $\log n$ for online path coloring in trees. We show that this algorithm can be forced to achieve $\Omega(\log(n))$ competitive ratio when applied to coloring paths in the line topology. However, this is far from the performance of the optimal 3-competitive algorithm by Kierstead and Trotter [9]. Our motivation is to understand this gap between performance of KT-algorithm and BL-algorithm for line topology. We present a simple online algorithm to solve the problem of online path coloring on caterpillars using at most $(5\omega - 3) < h(T)(5\omega - 3)$ colors, whereas the BL algorithm can be forced to use $\omega \log n$ colors by an adversary, details of which are given in Sect. 2.

For an arbitrary tree T, we define a hierarchical partition of the vertex set which we refer as the Hierarchical Path Partition (HPP). We associate a caterpillar with each part in the HPP, which results in a partition of the edge set of tree T. We call this edge partition a Hierarchical Caterpillar Partition (HCP) and we use an HCP in an online algorithm to color the path requests. In this online algorithm, we follow the template of Bartal and Leonardi [1]: each level in the HCP uses a distinct set of colors, and each coloring request is colored as a path coloring request at the highest level in which it intersects (has an edge in common) with a caterpillar. We denote by $h(T)$ the number of parts in an HPP with the minimum number of parts, and our algorithm uses at most $h(T)(5\omega-3)$ colors, where ω is the size of the maximum clique in the edge intersection graph of the given set of path coloring requests. Since ω is a lower bound on the number of colors to be used, our algorithm is an $O(h(T))$ competitive algorithm. It also gives us a refined understanding of the performance of the algorithm based on structure of tree T. To the best of our knowledge, the concept of hierarchical path partitioning is new and we believe it could be of significant interest in designing online algorithms for different problems on trees.

We also present an algorithm that computes an HPP with $h(T)$ parts in polynomial time. This algorithm also serves as a characterization of the optimum HPP. We then show that the optimum HPP has a subtree that *looks* like a complete binary tree of depth $h(T) - 1$. We refer to this as a *complete pseudo binary tree*. We use this subtree in conjunction with the lower bound argument due to Bartal and Leonardi [1] to show a competitive ratio lower bound of $\Omega(\frac{h(T)}{\log(4h(T))})$ for any deterministic online algorithm. In particular, our results nicely generalize the results of Bartal and Leonardi [1]- we design a $h(T)$ competitive online algorithm for path coloring in T, and also show a lower bound of $\Omega(\frac{h(T)}{\log(4h(T))})$ on the competitive ratio of any deterministic algorithm.

Definitions and Notation: We use standard graph theoretic concepts like *graph G*, *vertex set V(G)*, *edge set E(G)*, *degree $deg_G(v)$* of a vertex v, *neighborhood $N_G(v)$* of a vertex v, *diameter Δ*, *path $P = [v_1, v_2, \cdots, v_k]$* and *tree T* from the textbook by Douglas B. West [17]. The size of the largest clique in G is called its *clique number* and is denoted by $\omega(G)$. We use ω to denote the clique number of the edge intersection graph of input path requests of the underlying tree T. A caterpillar is a tree that has a dominating path. This dominating path is called the spine and an edge for which exactly one vertex is on the spine is called a hair. By definition all leaves other than the two on the spine are adjacent to a vertex on the spine. A *balanced tree separator* is a vertex whose removal splits the tree into multiple disjoint trees to form a forest such that each tree in that forest consists of at most $\frac{2}{3}n$ vertices. We denote path coloring requests by P and paths which are not path coloring requests by p.

Online Interval Coloring: Online path coloring in a tree T which is a path is the well-studied Online Interval Coloring problem. Kierstead and Trotter [9] gave a 3-competitive algorithm for the online interval coloring, which uses atmost $3\omega - 2$ colors where ω is the maximum number of pairwise intersecting intervals. They also showed a matching lower bound; that is, no deterministic online algorithm can achieve a competitive ratio better than 3. We refer to this online algorithm as the KT Algorithm.

2 Caterpillar Based Online Coloring of Paths in Trees

The main result in this section is our online algorithm for coloring paths in a tree T by partitioning the edges of T into parts, each of which is a set of vertex disjoint caterpillars. We show that this algorithm has a competitive ratio of $h(T)$, where $h(T)$ is a combinatorial parameter associated with T. The algorithm is based on our observation that the online path coloring algorithm (referred to as the BL Algorithm) due to Bartal and Leonardi [1] essentially maintains a partition of the edges of T, such that each part in the partition is a set of vertex disjoint stars - a special type of caterpillar.

A Worst Case Instance for the BL Algorithm: The first observation we make is based on the construction of a sequence of coloring requests in an n-vertex path to show that the BL algorithm has an $\Omega(\log n)$ competitive ratio. The BL algorithm has two phases: a preprocessing phase and a coloring phase. The preprocessing phase partitions the edges of T into $\log n$ levels. We refer to this partition as the BL partition. The preprocessing phase and the coloring phase are standard throughout all the algorithms we present, and the preprocessing phase depends only on the tree T and not on the path coloring requests.

Algorithm 1. BL

1: *Preprocessing Phase:* The first level L_0 consists of a single vertex s which is a balanced separator of T. Iteratively, for $i \geq 1$, the ith level L_i consists of balanced separators of all the subtrees in $T \setminus \bigcup_{0 \leq j \leq i-1} L_j$. In this way, vertices of T are partitioned into $\Theta(\log n)$ levels. The edge partition is obtained by associating with each vertex the set of incident edges whose other end point is at a higher level. Indeed, it is clear that each level is one set in a partition of the edges, and that the edges associated with a vertex at any level forms a star.

2: *Coloring Phase:* When a coloring request for a path P arrives, the algorithm first assigns a level identifier to it. This identifier is the minimum level number of a level that contains a vertex of P. Then, P is assigned the minimum color which is not assigned to any other previously colored path that has a common edge with P and has the same level identifier.

Theorem 1 (Bartal and Leonardi [1]). *The BL algorithm for online path coloring on a tree of n vertices uses at most $(2\omega - 1)\log n$ colors. Thus, the algorithm achieves a competitive ratio of $O(\log n)$.*

We now observe that BL algorithm has a competitive ratio of $\Omega(\log n)$ even if T is a path. On the other hand, the KT algorithm uses only $3\omega - 2$ colors to color paths from T if it is a path. We present an input instance generated by an adversary that forces the BL Algorithm to use $(\log n)OPT$ colors, where OPT is the number of colors in the optimal coloring. For an integer $k > 1$, consider the path T having $n = 2^k - 1$ vertices v_1 to v_{2^k-1}. Since there are $2^k - 1$ vertices, in the preprocessing phase, the vertices are partitioned into $L \geq k$ levels in any BL partition. Now, for each level $0 \leq l < L - 1$ in the BL partition, the adversary selects one edge of T with one end point at level l and the other end point at a level at least $l + 1$. Let these edges be $e_0, \ldots, e_{L-2}$. Further v be a vertex in level L (note that it has index $L - 1$) in the BL partition. In the path coloring sequence, there are $k(L - 1)$ paths consisting of exactly one edge and k paths consisting of exactly one vertex as follows: k paths consisting of e_0, followed by

k paths consisting e_1, and so on, ending with k paths consisting of e_{L-2} and, k single vertex paths consisting of v. Since the BL algorithm uses a distinct set of colors for each of the L levels, it uses at least $kL \geq (\log n)k = (\log n)OPT$.

Online Path Coloring in Caterpillars: The BL algorithm partitions the edges of T into $O(\log n)$ levels of stars. We explore if we can get a better competitive ratio by maintaining a set of caterpillars in each level. To achieve this, we first observe that paths in caterpillars can be colored by an online algorithm of competitive ratio 5. We refer to the algorithm below as Algorithm OPCC.

Algorithm 2. *OPCC- Online Path Coloring in Caterpillars*

1: *Preprocessing Phase:* In the preprocessing phase, we partition the edges of T into two sets: E_s is the set of all the edges on the spine and E_h is the set of all hairs.
2: *Coloring Phase:* We use two disjoint set of colors, C_s to color paths which have an edge in E_s and C_h to color paths whose edges are in E_h. When a coloring request for a path P arrives, it is colored as follows:
   Case 1: P contains an edge in E_s- In this case, we consider the subpath P' of P formed by the edges in E_s as an online path coloring request on the spine. P' is colored using the KT algorithm [9] (see Sect. 1.1)and the color of P' is the color given to P.
   Case 2: P does not contain any edge in E_s- In this case P must belong to some star in $E \setminus E_s = E_h$. P is greedily colored such that it gets a color different from paths colored earlier with which it shares an edge.

Lemma 1. *Let T be a caterpillar. Algorithm OPCC requires at most $5\omega - 3$ colors to color path coloring requests on T.*

Proof. The paths which fall into the first case are colored by the KT Algorithm [9] which uses at most $3\omega - 2$ colors. Secondly, the greedy algorithm on the paths colored in case 2 will use at most $2\omega - 1$ colors. The reason is that each such path has at most two edges, and if a path coloring request P cannot be colored by any of the already used $2\omega - 1$ colors, then one of the two edges in P is already in at least ω paths that have already been colored. This edge would then be common to $\omega + 1$ paths, contradicting the definition of ω. Hence, at most $5\omega - 3$ colors are required to color all input paths. $\square$

2.1 A New Online Path Coloring Algorithm for Trees

While the best competitive ratio for online path coloring in caterpillars is at most 5, we have shown that the BL algorithm has an $\Omega(\log n)$ competitive ratio when T is a caterpillar (because a simple path is a special caterpillar). We have also observed that for each tree T, the BL algorithm maintains a partition of the edge set into levels such that in each level, the edges form a set of vertex disjoint stars. We now present our algorithm where in the preprocessing phase, we maintain a partition of the edge set of T into levels such that in each level, the edges form a set of vertex disjoint caterpillars. In a way, this can be seen as a strengthening of the BL Algorithm. Then, by using the online coloring algorithm on caterpillars, we show that our online algorithm achieves a competitive ratio of $h(T)$, where $h(T)$ is defined as the number of levels in an optimal partitioning of the vertices that we call the Hierarchical Path Partition.

Hierarchical Path Partitioning: For a tree T, we call a partition $\{H_1, H_2 \ldots, H_h\}$ of $V(T)$ a *hierarchical path partitioning (HPP)* of T if the following properties hold:

P.1 Each set H_i induces a set of vertex disjoint paths in T. The paths in H_i are denoted by $\{p_{i,1}, p_{i,2}, \cdots, p_{i,l_i}\}$.

P.2 The sets are arranged hierarchically with one root set H_h. Note that at this place we have an important notational difference from Bartal and Leonardi [1] who use the level index 0 for the root.

P.3 Root set H_h contains a single path.

P.4 For each $i < h$, if p is a path in set H_i, then there is exactly one edge from exactly one of the vertices of path p to one vertex of a path $p' \in H_j$ where $j > i$. Further, this edge is incident on one of the end vertices of p. We refer to p' as the parent path of p.

P.5 For each $i > 1$, if p is a path in set H_i, then there are at least two edges from one endpoint of p to endpoints of paths p' and p'' in H_j where $j < i$. We refer to p', p'' as the children of p.

H_i in the partition is referred to as *level* i in the hierarchy. We use the HPP output by the following algorithm and the number of parts h output by this algorithm is referred to as $h(T)$.

Algorithm 3. OHPP: Optimum Hierarchical Path Partitioning

We define a sequence of non-empty subtrees $\{T_i\}$ of tree T where $T_1 = T$. The vertex set of T_i is $V(T_{i-1}) \setminus H_{i-1}$. For some $h \geq 1$, if T_h is a path, then H_h consists of the single path, and the algorithm stops and outputs $\{H_1, H_2 \ldots, H_h\}$. For each $i \geq 1$, $H_i = \{p_{i,1}, p_{i,2} \cdots, p_{i,l_i}\}$ is a set of vertex disjoint paths, such that for each $p_{i,j} \in H_i$, one endpoint is a leaf in T_i, the other endpoint of $p_{i,j}$ has a neighbor in T_i whose degree in T_i is at least 3, and all other vertices in $p_{i,j}$ have degree 2 in T_i. Clearly, $T_{i+1} = T_i \setminus H_i$ induces a subtree of T_i. The algorithm is illustrated in Fig. 1.

From an HPP we naturally obtain a unique hierarchical partition of the edge set of T into vertex disjoint sets of caterpillars as follows: For each path $p_{i,j} \in H_i$, $1 \leq i \leq h$, we associate a caterpillar $c_{i,j}$ by taking $p_{i,j}$ to be the spine and every other edge e incident on vertices of $p_{i,j}$ such that other vertex of e is on a path in H_k for some $k < i$. This produces a hierarchical partitioning of the edge set of T into caterpillars called *Hierarchical Caterpillar Partitioning (HCP)* of T. Let $C_1, \ldots, C_h$ denote this family of sets of caterpillars corresponding to HPP $H_1, \ldots, H_h$.

Online Path Coloring Using a HCP: We now describe our Algorithm OPCT to color path requests in a given tree T. As in the BL Algorithm 1, we have a preprocessing phase and a coloring phase.

Algorithm 4. *OPCT- Online Path Coloring in Trees*

1: *Preprocessing phase:* Compute the natural (as described above) HCP $\{C_1, \ldots, C_h\}$ of the given tree T from an Optimum HPP computed using Algorithm OHPP described in Sect. 3. Here $h = h(T)$.

2: *Coloring Phase:* For each level in the HCP, a distinct set of colors is used. When a coloring request for a path P arrives, it is assigned a level number q which is the maximum among all levels with which path P intersects at some edge in the level. Let q be the largest level such that there is a caterpillar $c_{q,j} \in C_q$ with which P intersects at an edge in $c_{q,j}$ (In the proof below, we use q to denote this level associated with a coloring request P. For example, for a path P_1, q_1 is used to denote this level). Then, the subpath P' of P consisting of edges of $E(P) \cap E(c_{p,j})$ is considered as a path coloring request in $c_{p,j}$ (In the proof below, we use P' to denote the coloring request in $c_{q,j}$ associated with P. For example, for a path P_1, P_1' is used to denote the coloring request in $c_{q_1,j}$). It is colored using Algorithm OPCC, and the color given to P' is taken to be the color of P.

Theorem 2. *Let T be a tree, then online Algorithm OPCT requires at most $h(5\omega - 3)$ colors to compute a valid coloring on an online request sequence of paths from tree T such that any edge is present in at most ω paths.*

Proof. To prove that the coloring is valid, we first need to observe that any path P intersects at an edge with exactly one caterpillar in C_q. We know from Lemma 1 that the path coloring request sequence in each caterpillar in each level of the HCP uses at most $5\omega - 3$ colors. Since the HCP is an induced set of vertex disjoint caterpillars (therefore, edge disjoint), it follows that the path coloring request sequence to each level of the HCP uses at most $5\omega - 3$ colors. Since each level uses a distinct set of colors, it follows that Algorithm OPCT uses at most $h(5\omega - 3)$ colors. Let P_1 and P_2 be two path coloring requests that share an edge in T. We show that the color given to P_1 and P_2 are distinct. If q_1 and q_2 are different, then P_1' and P_2' get different colors and therefore P_1 and P_2 get different colors. In the case when $q_1 = q_2$, since P_1 and P_2 intersect at an edge, it follows that they intersect with the same caterpillar $c_{q_1,j} \in C_{q_1}$. Further, it is easy to see that they also share a common edge in $c_{q_1,j}$. Thus it follows that P_1' and P_2' get different colors, and consequently P_1 and P_2 get different colors. Hence the theorem.   □

3   A Lower Bound for Deterministic Online Algorithms Using $h(T)$

In this section we prove a lower bound on the competitive ratio of deterministic online algorithms as a function of $h(T)$. We start by illustrating in Fig. 1 a HPP computed by Algorithm 3 and observing some bounds on $h(T)$.

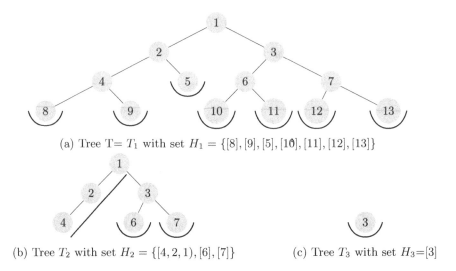

(a) Tree T= T_1 with set $H_1 = \{[8],[9],[5],[10],[11],[12],[13]\}$

(b) Tree T_2 with set $H_2 = \{[4,2,1],[6],[7]\}$ (c) Tree T_3 with set $H_3=[3]$

Fig. 1. Tree T with partition $\{H_1, H_2, H_3\}$. (a) T_1 with H_1, (b) T_2 with H_2, (c) T_3 with H_3

Some Bounds on $h(T)$: We observe that, for a tree T having L leaves, $h(T) \leq \log_2 L$, as the number of leaves at each level reduce by a factor of at least 2. Formally, let l_i be the number of leaves in T_i. Then, $l_{i+1} \leq \frac{l_i}{2}$. Similarly, the diameter of T_{i+1} is at most diameter of T_i minus 2. Therefore, $h(T) \leq \frac{\Delta(T)}{2}$, where Δ is the diameter.

We next show that the HPP computed by Algorithm OHPP can be used by an adversary to ensure that any deterministic online algorithm has a *bad* competitive ratio.

3.1   A Lower Bound Based on $h(T)$

To prove our lower bound on deterministic online algorithms, we use the lower bound technique of Bartal and Leonardi [1] on path coloring requests on a complete binary tree of n nodes. To use this, we show that we can perform their adversarial lower bound argument on a complete pseudo binary tree which we show is present in the output of Algorithm $OHPP$. Intuitively, the subtree of T that we take will turn out to be a tree that looks like a complete binary tree whose vertices correspond to paths in which each internal vertex has degree 2. We now describe this Complete Pseudo Binary Tree. Consider the partitioning obtained after applying the algorithm $OHPP$ to T. We know that the output of Algorithm $OHPP$ is an optimum HPP $H_1, \cdots, H_h$. We call a family $\{S_1, S_2 \ldots, S_h\}$ of sets of paths a *Complete Pseudo Binary Tree* if the following properties hold.

- For each $1 \leq i \leq h$, $S_i \subseteq H_i$, S_i consists of 2^{h-i} paths, and $S_h = H_h$.
- Let v be one end point on the path $p \in H_h$. Then, S_{h-1} consists of exactly 2 paths, say p' and p'' such that v is adjacent to one end point each of p' and p''.

- For each $1 \leq i \leq h-1$, in every path p in S_{i+1}, there are two edges from one end point of p to the end points of two paths in S_i.
- For each $1 \leq i \leq h-1$, in every path p in S_i there is an edge from one end point of p to the end point of a path in S_{i+1}.

Lemma 2. *Let T be a tree, $h = h(T)$, and let $H_1, \ldots, H_h$ be the optimal HPP obtained from Algorithm OHPP. T has a subtree which is a complete pseudo binary tree of height h.*

Proof. We prove our claim by induction on the number of levels $h(T)$ in the partition of T produced by our algorithm. Let $h = h(T)$. Let $L(h)$ be the statement that there exists a complete pseudo binary tree in the hierarchical path partitioning of at most h levels. The base case is the statement $L(1)$. Then, $L(1)$ is true because there does exist a complete pseudo binary tree having 1 pseudo-node. Suppose the statement $L(h)$ is true. We now prove the statement $L(h+1)$. Let $\{H_1, \ldots, H_{h+1}\}$ be the hierarchical path partitioning having $h+1$ levels output by Algorithm OHPP. Now, consider the tree $T_2 = T \setminus V(H_1)$ with HPP $\{H_2, \ldots, H_{h+1}\}$ obtained after removing all the paths in H_1. By the induction hypothesis, there exists a complete pseudo binary tree for T_2 in $H_2, \ldots, H_{h+1}$. Let this complete pseudo binary tree be $S_2, \ldots, S_{h+1}$. Now, consider the paths in S_2. For each $p \in H_2$, let l_p be the leaf endpoint of p in T_2. By the definition of an HPP, l_p is incident on at least two edges, for each of which the second vertex (the one different from l_p) is an endpoint of a path in H_1. We construct $S_1 \subseteq H_1$ by taking any two such paths for l_p for each $p \in S_2$. Now, $S_1, \ldots, S_{h+1}$ is a complete pseudo binary tree for T. Thus, there exists a complete pseudo binary tree of T in the output of Algorithm OHPP with at most $h+1$ levels. Hence the lemma.   □

We now use this complete pseudo binary tree to get our lower bound on deterministic online algorithms for path coloring on a tree T. As mentioned before, we essentially plug this complete pseudo binary tree into the lower bound argument of Bartal and Leonardi [1].

Theorem 3. *Let T be a tree. Then any deterministic online path coloring algorithm has a competitive ratio of $\Omega(\frac{h(T)}{\log 4h(T)})$.*

Description of the Adversarial Path Coloring Request Sequence: Before we present a proof of this theorem, we describe the sequence of requests presented by an adversary to a deterministic online path coloring algorithm. The lower bound is against a deterministic online algorithm and is established by an adversary by using the complete pseudo binary tree $S = \{S_1, S_2 \ldots, S_h\}$ contained in the optimal HPP $\{H_1, H_2 \ldots, H_h\}$ output by Algorithm OHPP. This complete pseudo binary tree has $h(T)$ levels, with S_h containing the root pseudo-node, and S_1 containing all the leaf pseudo-nodes. Each pseudo-node here corresponds to a path in S. We consider a complete binary tree T' of depth $h(T) - 1$ where there is a bijective correspondence between the pseudo-nodes and the vertices of T', that respects the parent-child relationships between the paths in the family S.

For example, the root of T' corresponds to the path in S_h, and the leaves of T' corresponds to the paths in S_1. A path coloring request in T' is converted to a path coloring request in T by *expanding* the pseudo-nodes into the corresponding paths in the family S. In the following let ρ denote the best competitive ratio possible by any deterministic online algorithm.

We now present the request sequence generated by Bartal and Leonardi [1] on T'. The description is identical to the description in [1] except in the Lemma 3 where we reason about the path coloring requests in T obtained from the requests in T' as described below. The sequence of requests for coloring paths is generated in stages. Maintain the following invariant at the end of any stage $i \geq 0$: There exists a set C_i of i colors, a level $\bar{l}_i$, $l_i \leq \bar{l}_i \leq h(T) - 1 - i$, such that there are at least $r_i = \frac{2^{h(T)-1}}{(8\rho h(T))^i}$ pairs of paths with the following properties:

1. Each pair is formed by the two paths from two leaves in T' to their least common ancestor (LCA) at level $\bar{l}_i$.
2. Each vertex of level $\bar{l}_i$ in T', is the LCA of at most one pair of paths.
3. For any path in the r_i pairs of paths and for any color $c \in C_i$, there is one edge in the path included in a path coloring request with color c.
4. In T', any edge of a path is included in at most one request.

At stage 0, $\bar{l}_i = l_0 = h(T) - 1$ and $C_0 = \phi$. We associate a set of $r_0 = 2^{h(T)-1}$ pairs of empty paths, two with each leaf, with both endpoints equal to leaf itself. No path coloring requests are presented. Hence all 4 properties trivially hold. At stage $i + 1$, r_i new path coloring requests are presented, one for each pair of paths. Let u_1, u_2 be the two leaves that are endpoints of the two paths of a pair, and let $LCA(u_1, u_2)$ be the LCA at level $\bar{l}_i$ of these two leaves. Let v be the direct ancestor of $LCA(u_1, u_2)$. For each pair of paths we present a path coloring request having as endpoints one of the two leaves, say u_1, and v. The online algorithm must color the set of path coloring requests presented at stage $i + 1$. Clearly, due to the 4 invariants being respected at the beginning of stage $i + 1$, any color in C_i cannot be used for these path coloring requests. These path coloring requests on T' are converted to path coloring requests in the complete pseudo binary tree. We will show in Lemma 3 that optimal number of colors at any stage of the sequence is at most 2 in the complete pseudo binary tree. Hence, in order for the online algorithm to be ρ competitive, it must use less than 2ρ colors for this set of path coloring requests. Therefore, by the pigeonhole principle there must be a set of path coloring requests R_i of cardinality at least $\frac{r_i}{2\rho}$ assigned with the same color. Let us call this color c_{i+1} and let $C_{i+1} = C_i \cup \{c_{i+1}\}$. In the following, we concentrate on this set of path coloring requests. As shown in Bartal and Leonardi [1] we first identify a set of paths in R_i that satisfy condition 3 and 4. Recall that each path coloring request in R_i is from a leaf u_1 to a node v, is constructed from some level i pair of paths from the leaves u_1 and u_2 to $LCA(u_1, u_2)$, which is a child of v. Moreover, for any pair only one path coloring request in R_i is constructed. Therefore, for any path coloring request in R_i, conditions 3 and 4 are satisfied at stage $i + 1$ for the path connecting the leaf u_2 to v or any ancestor of v. In fact, at most one path coloring request includes any edge from u_2 to $LCA(u_1, u_2)$ and any color $c \in C_i$ is associated with a path coloring request that crosses only one edge in the path, using the invariant for level i. Moreover, the edge from $LCA(u_1, u_2)$ to v is associated with only one path coloring request with color c_{i+1}. We thus have a set of $\frac{r_i}{2\rho}$ paths satisfying conditions 3 and 4 of the invariant. We call this set of paths P_{i+1}.

The level l_{i+1} is now selected such that $2^{l_{i+1}+1} \leq \frac{r_i}{2\rho}$. We derive from P_{i+1} a new set P'_{i+1} as follows: we consider each path in P_{i+1} according to its ancestor in level l_{i+1}. If a vertex in level l_{i+1} is an ancestor of an odd number of paths we exclude one of these paths. Since the number of vertices of level l_{i+1} is at most $\frac{1}{2}\frac{r_i}{2\rho}$, the cardinality of P'_{i+1} is at least $\frac{1}{2}\frac{r_i}{2\rho}$.

We now scan paths in P'_{i+1} from left to right, following the order of the leaves that are endpoints of those paths. We associate each pair of successive leaves with their LCA, that is a vertex of level between l_{i+1} and $h(T) - i - 1$. Again, as shown in [1] we know that each vertex in a binary tree is the LCA of at most one pair of successive leaves.

Finally, let $\overline{l_{i+1}}$ be a level between $h(T) - i - 1$ and l_{i+1}, achieving the maximum cardinality set of pairs of successive paths that have LCA at that level. We define the set of pairs of paths for the stage $i + 1$ to be the set of pairs of successive paths that have LCA at level $\overline{l_{i+1}}$. Since the number of levels is $h(T)$, it follows that the number of pairs at stage $i + 1$ is atleast $\frac{1}{4}\frac{r_i}{2\rho h(T)} = \frac{2^{h(T)-1}}{(8\rho h(T))^{i+1}} = r_{i+1}$. From the above construction, it follows that both conditions 1 and 2 hold for this set of pairs. Therefore, the four invariants are satisfied at the beginning of stage $i + 1$.

We now come to the crucial and only modification to the argument of Bartal and Leonardi [1].

Lemma 3. *The optimal solution in the complete pseudo binary tree for the path coloring sequence described above uses at most 2 colors.*

Proof. The proof is obtained by the fact that in T' each edge is included in at most 2 path coloring requests, and that all such path coloring requests are directed from a leaf to an ancestor. Further, at each vertex in the tree, at most one edge to a child is present in the path coloring requests. Therefore, when these requests are considered as requests in the complete pseudo binary tree, each edge is present in at most two path coloring requests. Therefore, the paths can be colored offline with 2 colors. □

Proof of Theorem 3: Let ρ be the competitive ratio of the best deterministic online algorithm for coloring paths in tree T. Clearly, $\rho \leq \frac{h(T)(3\omega-2)}{\omega}$. Therefore, $\rho \leq 2h(T)$. The online algorithm uses at least i colors after i stages of the construction above. Hence by Lemma 3, the competitive ratio $\rho \geq \frac{i}{2}$. The lower bound on ρ is thus obtained by computing the maximum number of stages in the sequence. To carry out the sequence, we require that $l_i = h(T) - 1 - i\log(8\rho h(T)) \geq 1$. Since $\rho \geq \frac{i}{2}$, we get $\rho \geq \frac{h(T)-2}{2\log(8\rho h(T))}$. Since we know that $\rho \leq 2h(T)$, it follows that $\rho \geq \frac{h(T)-2}{2\log(16h(T)^2)}$. Therefore, ρ is $\Omega(\frac{h(T)}{\log 4h(T)})$. Hence the theorem. □

Acknowledgments. We would like to thank Krithika Ramaswamy, Dhannya S.M., and Stefano Leonardi for helpful suggestions.

References

1. Bartal, Y., Leonardi, S.: On-line routing in all-optical networks. Theor. Comput. Sci. **221**(1–2), 19–39 (1999)
2. Erlebach, T., Jansen, K., Elvezia, C.: The complexity of path coloring and call scheduling. Theoret. Comput. Sci. **255**, 2001 (2000)
3. Gavril, F.: Algorithms for minimum coloring, maximum clique, minimum covering by cliques, and maximum independent set of a chordal graph. SIAM J. Comput. **1**(2), 180–187 (1972)
4. Gavril, F.: The intersection graphs of subtrees in trees are exactly the chordal graphs. J. Comb. Theor. Ser. B **16**(1), 47–56 (1974)
5. Gavril, F.: A recognition algorithm for the intersection graphs of paths in trees. Discrete Math. **23**(3), 211–227 (1978)
6. Golumbic, M.C., Jamison, R.E.: Edge and vertex intersection of paths in a tree. Discrete Math. **55**(2), 151–159 (1985)
7. Golumbic, M.C., Jamison, R.E.: The edge intersection graphs of paths in a tree. J. Comb. Theo. Ser. B **38**(1), 8–22 (1985)
8. Kaklamanis, C., Persiano, P.: Efficient wavelength routing on directed fiber trees. In: Diaz, J., Serna, M. (eds.) ESA 1996. LNCS, vol. 1136, pp. 460–470. Springer, Heidelberg (1996). doi:10.1007/3-540-61680-2_75
9. Kierstead, H.A., Trotter, W.T.: An extremal problem in recursive combinatorics. Congressus Numerantium **33**(143–153), 98 (1981)
10. Kumar, V., Schwabe, E.J.: Improved access to optical bandwidth in trees. In Proceedings of SODA 1997, pp. 437–444 (1997)
11. Mihail, M., Kaklamanis, C., Rao, S.: Efficient access to optical bandwidth. In: Proceedings of the 36th Annual Symposium on Foundations of Computer Science, FOCS 1995, p. 548. IEEE Computer Society, Washington, DC (1995)
12. Narayanaswamy, N.S., Subhash, R.: Babu. A note on first-fit coloring of interval graphs. Order **25**(1), 49–53 (2008)
13. Pemmaraju, S.V., Raman, R., Varadarajan, K.R.: Buffer minimization using max-coloring. In: Ian Munro, J. (ed.) SODA, pp. 562–571. SIAM (2004)
14. Raghavan, P., Upfal, E.: Efficient routing in all-optical networks. In: Proceedings of the Twenty-sixth Annual ACM Symposium on Theory of Computing, STOC 1994, pp. 134–143. ACM, New York (1994)
15. Takai, H., Kanatani, T., Matsubayashi, A.: Path coloring on binary caterpillars. IEICE Trans. **89–D**(6), 1906–1913 (2006)
16. Tarjan, R.E.: Decomposition by clique separators. Discrete Math. **55**(2), 221–232 (1985)
17. West, D.B.: Introduction to Graph Theory, 2nd edn. Prentice Hall, Upper Saddle River, NJ (2000)

Resource Allocation Games with Multiple Resource Classes

Roy B. Ofer[(✉)] and Tami Tamir[(✉)]

School of Computer Science, The Interdisciplinary Center, Herzliya, Israel
royofr@gmail.com, tami@idc.ac.il

Abstract. We define and study a resource-allocation game, arising in Media on Demand (MoD) systems where users correspond to self-interested players who choose a MoD server. A server provides both storage and broadcasting needs. Accordingly, the user's cost function encompasses both positive and negative congestion effects.

A system in our model consists of m identical servers and n users. Each user is associated with a type (class) and should be serviced by a single server. Each user generates one unit of load on the server it is assigned to. The load on the server constitutes one component of the user's cost. In addition, the service requires an access to an additional resource whose activation-cost is equally shared by all the users *of the same class* that are assigned to the same server. In MoD systems, the bandwidth required for transmitting a certain media-file corresponds to one unit of load. The storage cost of a media-file on a server is shared by the users requiring its transmission that are serviced by the server.

We provide results with respect to equilibrium existence, computation, convergence and quality. We show that a pure Nash Equilibrium (NE) always exists and best-response dynamics converge in polynomial time. The equilibrium inefficiency is analyzed with respect to the objective of minimizing the maximal cost. We prove that the Price of Anarchy is bounded by m and by the size of the smallest class and that these bounds are tight and almost tight, respectively. For the Price of Stability we show an upper bound of 2, and a lower bound of $2 - \frac{1}{m}$. The upper bound is proved by introducing an efficient 2-approximation algorithm for calculating a NE. For two servers we show a tight bound of $\frac{3}{2}$.

1 Introduction

Resource allocation problems consider scenarios in which tasks or clients have to be assigned to resources under a set of constraints. Resource allocation applications exist in a variety of fields ranging from production planning to operating systems. Game theoretic considerations have been studied in many resource allocation problems. The game theoretic view assumes that users have strategic considerations acting to maximize their own utility, rather than optimizing a

---

A brief-announcement introducing this work was presented in the 8th International Symposium on Algorithmic Game Theory (SAGT), 2015.

© Springer International Publishing AG 2017
K. Jansen and M. Mastrolilli (Eds.): WAOA 2016, LNCS 10138, pp. 155–169, 2017.
DOI: 10.1007/978-3-319-51741-4_13

global objective. In resource allocation problems, this means that users *choose* which resources to use rather than being assigned to resources by a centralized designer. Media streaming is among the most popular services provided over the Internet. The lack of a central authority that controls the users, motivates the analysis of Media on Demand (MoD) services using game theoretic concepts.

Two main approaches exist with respect to the cost function associated with the usage of a resource. One approach considers congestion games in which user's cost increases with the load on the resource. The other approach considers cost sharing games in which users share the activation-cost of a resource, and thus, user's cost decreases with the load on the resource. Feldman and Tamir introduced and studied a model in which both considerations apply [4]. In this work we generalize this model further and study games corresponding to systems in which resources have both positive and negative congestion effects, and different users may require different resources. Our work is motivated by Media-on-Demand systems, in which the above cost scheme applies.

A system in our model consists of a set of identical servers. Each user of the system is associated with a type (class) and should be serviced by a single server. Every user generates one unit of load on the server it is assigned to. In addition, the service requires an access to an additional resource whose activation-cost is equally shared by all the users *of the same type* that are assigned to the server.

A configuration of the system is characterized by an allocation of users to servers. The cost of a user in a given allocation is the sum of two components: the load-cost determined by the total load on his server, and his share in the class activation-cost.

A pure Nash equilibrium (NE) is a configuration in which no individual player can migrate and reduce his cost. We study the multi-class resource model with respect to NE existence, calculation and efficiency. When considering equilibrium inefficiency we use the standard measures of price of anarchy (PoA) [7] and price of stability (PoS) [2]. For the PoA and PoS measures we use an egalitarian objective function, i.e., we measure the maximal cost among users compared with the maximal cost in an optimal allocation. In addition to the theoretical analysis of this model, we present efficient algorithms for finding good stable solutions. The algorithms combine load-balancing ideas used in packing algorithms, such as element-grouping and handling the elements in decreasing-size order, together with ideas used in algorithmic game theory, such as performing a sequence of improving steps in a specific, supervised, order.

Applications: There are several real-world systems that fit the above multi-class resource allocation scenario. In particular, our study is motivated by media-on-demand (MoD) systems. A MoD system (see, e.g., [13,17]) consists of a large database of media files and a set of servers. The servers provide both storage and broadcasting needs. Each client specifies a media stream request and receives the stream via one of the servers. The server's bandwidth corresponds to the load resource – each client generates one unit of load on the server. The media-file specifies the client's class. Each media-file (class) has an activation-cost reflecting the cost of copying the media file from the central database, and storing it in the server's local memory. The server's bandwidth (load) is distributed among all its clients, while the class activation-cost is shared among all clients requiring the same media file stream.

Another example is infrastructure-as-a-service (IAAS) in cloud computing. IAAS (see e.g. [10]) is a cloud computing service model which offers computers, either physical or virtual machines. Each client has a task that has to be performed on a machine. In IAAS system, each machine acts as a server. The machine's network bandwidth corresponds to the load resource and the required software installation for the client's task specifies the class. The load on the virtual machine affects all the machine's clients, while the software installation cost is shared among all clients requiring it.

Production planning is another example of a multi-class resource allocation application, arising in computer systems and in many other areas. Consider a set of machines, each having a limited capacity of some physical resource (e.g. quantity of production materials). In addition, hardware specifications allow each machine to produce items of different types, each associated with some configuration set-up or training. The quality of service reduces with the total congestion on the resource. The configuration set-up cost is required for every class on every machine.

1.1   Model and Preliminaries

An instance of the multi-class resource allocation game is defined by a tuple $G = \langle I, M, A, U \rangle$, where I is a set of players, M is a set of servers and A is a set of classes. Let $n = |I|$ and $m = |M|$. Each player belongs to a single class from A, thus, $I = I_1 \cup I_2 \cdots \cup I_{|A|}$, where all players from I_k belong to class k. For $i \in I$, let $a_i \in A$ denote the class to which player i belongs. The parameter $U \in \mathbb{R}^+$ is the *class activation-cost*, which is assumed to be uniform for all classes.

An *allocation* of players to servers is a function $f : I \to M$. Given an allocation, the *load* on server j, denoted by $L_j(f)$, is the number of players assigned to j. We denote by $L_{j,k}(f)$ the number of players from I_k assigned to j. When clear in the context we omit f and use L_j and $L_{j,k}$, respectively.

The cost of a player i in an allocation f consists of two components: the load on the server the player is assigned to, and the player's share in the class activation-cost. The class activation-cost is shared evenly among the players from this class serviced by the server. Formally, $c_f(i) = L_{f(i)} + \frac{U}{L_{f(i),a_i}}$.

A *step* by a player i with respect to an allocation f is a unilateral deviation of i, i.e., a change of f to f' such that $\forall_{\ell \neq i} f'(\ell) = f(\ell)$ and $f'(i) \neq f(i)$. An *improving step* of player i with respect to an allocation f is a step which reduces the player's cost, that is, $c_{f'}(i) < c_f(i)$. An allocation f is said to be a *Pure Nash Equilibrium* (NE) if no player has an improving step, i.e., for each player i and for every allocation f' such that $\forall_{\ell \neq i} f'(\ell) = f(\ell)$ it holds $c_f(i) \leq c_{f'}(i)$.

Best-Response Dynamics (BRD) is a local search method where in each step some player is chosen and plays its best improving step, given the strategies of the other players.

It is well known that decentralized decision-making may lead to sub-optimal solutions from the point of view of society as a whole. We quantify the inefficiency incurred due to self-interested behavior according to the PoA and PoS measures. The PoA is the worst-case inefficiency of a NE, while the PoS measures the best-case inefficiency of a NE. Formally, let $\mathcal{G}$ be a family of games, and let $G \in \mathcal{G}$ be some game in this family. Let $NE(G)$ be the set of Nash equilibria of the game G, let $val(f)$ be the social cost of a NE f with respect to some objective function, and let $OPT(G)$ be the value of an optimal solution. If $NE(G) \neq \emptyset$, then $PoA(G) = \max_{f \in NE(G)} \frac{val(f)}{OPT(G)}$, and $PoA(\mathcal{G}) = Sup_{G \in \mathcal{G}} PoA(G)$. Similarly, $PoS(G) = \min_{f \in NE(G)} \frac{val(f)}{OPT(G)}$, and $PoS(\mathcal{G}) = Sup_{G \in \mathcal{G}} PoS(G)$.

In this paper, we evaluate the performance of a solution with respect to the objective of minimizing the maximal cost among the players; that is, given an allocation f, the social cost of f is given by $c_{max}(f) = \max_{i \in I} c_f(i)$.

1.2   Related Work

The study of resource allocation games with multiple resource classes combines challenges arising in the two classical problems of multi-dimensional packing and resource-sharing games. Class-constrained multiple knapsack *(CCMK)* [13,14] is the variant of a centralized packing problem closest to our model. In CCMK each item has a type, a size and a value. Each knapsack has in addition to its size, a number of compartments which define the number of different item types it can contain. The optimization goal in CCMK is to maximize the total value of items packed into the knapsacks. The problem is NP-hard even with unit size and unit profit items. In our game, as in [13], all items have unit size. The main difference between the models is that servers in our game have no limited capacity, thus a placement that packs all the items always exists. The load-component in our cost-function provides the incentive to avoid highly loaded servers and to balance the load among the servers.

In cost-sharing games, a possibly unlimited amount of resources is available. The activation of a resource is associated with a cost which is shared among the players using it. A well-studied cost sharing game is network design. Nash equilibrium always exists in network design games and the price of stability with respect to the total-cost objective function is $H(k)$, where k is the number of players and H is the harmonic function [1]. In cost sharing games, congestion has a positive effect, and players have an incentive to use resources that are used by others. Other related work deal with congestion games, in which congestion has a negative effect, and players wish to avoid loaded resources. In congestion games, the cost of using a resource increases with the load on it. Congestion games were first introduced in [11], and arise naturally in network routing (see e.g. [12]), and job-scheduling [16].

In [4], Feldman and Tamir studied a model incorporating both positive and negative congestion effects. In their model, a job-scheduling setting with unlimited set of identical machines is studied. Each job j has a length p_j and each machine has a fixed activation cost U. The set of players corresponds to the set of individual jobs and the action space of each player j is the set of machines. The cost function of job j in a given schedule is composed of the load on the job's machine and the job's share in the machine's activation cost. For the uniform sharing rule in which the machine's activation cost is uniformly shared between the jobs allocated to it, a NE may not exists. For the proportional sharing rule in which the share of a job in the machine's activation cost is proportional to its length, the price of anarchy with respect to the makespan can be arbitrarily high. The price of stability is tightly bounded by 5/4. This model of conflicting congestion effects was studied further in [3], where equilibrium inefficiency was studied with respect to the total-cost objective, and in [5,8], where closer analysis of the PoA and PoS is provided. In this work, we generalize the model of conflicting congestion effects, by allowing several resources on a single server. This generalization provides one additional step in modeling real-world systems using game theoretical tools.

1.3   Our Results

We provide answers to the basic questions regarding resource allocation games with multiple resource classes. Namely, equilibrium existence, convergence, calculation and efficiency. We present polynomial-time algorithms for calculating a stable solution whose cost almost matches the bound for the PoS.

We prove that a NE exists for any instance of the game by presenting an exact potential function for the game. By analyzing this function we conclude that any application of better-response dynamics converges to a NE within time $O(n^4)$. The equilibrium inefficiency is analyzed with respect to the objective of minimizing the maximal cost among the players. We first provide several lower-bounds on the optimal solution, and then combine them to present a tight bound of m for the PoA. An additional almost tight bound depends on the size of a least popular class. Let $\theta = \min_{1 \leq k \leq |A|} |I_k|$. We show that $PoA \leq \theta + 1$, and a game for which $PoA \geq \theta - \epsilon$ exists.

We show that for any number of servers, there exists a game for which the PoS is $2 - \frac{1}{m}$. This lower bound is almost matched - we present a polynomial time algorithm that constructs a NE with max-cost at most twice the optimum. For two servers, we present a matching upper bound: that is - a polynomial time algorithm that constructs a NE with max-cost at most 3/2 times the optimum.

Our algorithms for finding a good stable assignment are based on two new methods:

1. While all the players create the same unit-load on the servers, our algorithms group the players into sets, based on their classes. An initial assignment is found by considering these sets as an instance of a multiple-knapsack packing problem with arbitrary-size elements. This method enables analysis of the assignment using known packing techniques and their properties.
2. The stabilization phase that follows the initial assignment consists of iterations in which the algorithm may reassign complete sets of players, or perform a *supervised* sequence of improving steps. The sequence is initiated by

one player i, and is then limited to players of i's class who may benefit from following i by performing exactly the same migration. Analyzing the configuration after each improving step is complex; however, it is possible to analyze the effect of each supervised sequence of improving steps on the potential function and to bound the cost of an assignment derived by this method.

It is interesting to compare our results with the model studied in [4], in which all users belong to a single class, and the number of servers is unlimited. In our model, it is not relevant to study instances with unlimited number of servers, since players from different classes have only negative effect on the cost of each other, thus, different classes will never share a server, and the problem reduces to a single-class problem. The PoA is not bounded by a constant in both models, however, in our model it is bounded by m – the number of servers and $\theta + 1$ - where θ is the size of the smallest class, while in [4] it is bounded by $\frac{1+U}{2\sqrt{U}}$ – which is a function of the class (machine) activation-cost.

A tight bound of $\frac{5}{4}$ for the PoS is shown in [4]. Our bound on the PoS implies that increasing the number of classes from 1 to arbitrary $|A|$ only slightly increases the PoS - from $\frac{5}{4}$ to 2. Thus, in both models the PoS is a relatively small constant.

Due to space constraints, some proofs as well as the $\frac{3}{2}$-approximation algorithm for two servers are omitted from this extended abstract.

2   Equilibrium Existence and BRD Convergence

We show that the multi-class resource allocation game is a *potential game* [9]. This implies that a series of improving steps always converges to a NE. Given an allocation f, consider the following potential function,

$$\Phi(f) = \sum_{1 \le j \le m} U \cdot (H_{L_{j,1}(f)} + H_{L_{j,2}(f)} + \ldots + H_{L_{j,|A|}(f)}) + \frac{L_j(f)^2}{2}, \quad (1)$$

where H_k is the kth harmonic number, that is, $H_0 = 0$, and $H_k = 1 + \frac{1}{2} + \ldots + \frac{1}{k}$.

Theorem 1. $\Phi(f)$ *is an exact potential function.*

Thus, BRD converges and a NE exists. Next, we show that BRD converges to a NE in polynomial time. Specifically,

Theorem 2. *For every instance G, BRD converges to a NE within $O(n^4)$ steps.*

Proof. Consider the potential function defined in (1), since $H_k \le k$, and for all $1 \le j \le m$ and $1 \le i \le |A|$, $L_{j,i} \le L_j$, the left addend of the sum can be bounded as follows,

$$\sum_{1 \le j \le m} U \cdot (H_{L_{j,1}(f)} + H_{L_{j,2}(f)} + \ldots + H_{L_{j,|A|}(f)}) \le \sum_{1 \le j \le m} U \cdot L_j = U \cdot n.$$

The right addend of the potential function is trivially bounded by $\frac{n^2}{2}$ and we conclude that for all f, $\Phi(f) \le U \cdot n + \frac{n^2}{2}$. Consider an improving step by some player i. Since the potential function is an exact potential function, the difference in the potential is exactly the improvement in i's cost. That is $\Delta\Phi = c_f(i) - c_{f'}(i) = \Delta c^\ell(i) + \Delta c^s(i)$. The difference in the load is an integer while the difference in the activation-cost is $\frac{U}{L_{f(i),a}(f)+1} - \frac{U}{L_{f(i),a}(f)}$ where a is the class of i. Since $L_{j,a}$ is an integer and $L_{j,a} \le n$ for all a, j, the denominator of the activation-cost diff is at most $n(n-1)$. Thus, an improving step reduces the potential by at least $\frac{1}{n(n-1)}$, that is, $\Delta\Phi \ge \frac{1}{n(n-1)}$. Since the potential is always positive, BRD converges in at most $\frac{max_f \Phi(f)}{min \Delta\Phi} = \frac{O(n^2)}{\Omega(\frac{1}{n^2})} = O(n^4)$ steps. $\qquad \square$

3 Equilibrium Inefficiency - Price of Anarchy

In this section we study the inefficiency caused due to strategic behavior, as quantified by the Price of Anarchy (PoA). We evaluate the performance of a solution with respect to the objective of minimizing the highest cost among all the players; that is, given an allocation f, the social cost of f is given by $c_{max}(f) = \max_{i \in I} c_f(i)$. For a server j, define the cost of j as the maximal cost among players allocated to j. That is, $c_f(j) = \max_{f(i)=j} c_f(i)$. Let OPT denote the maximal cost of a player in an optimal assignment minimizing the maximal cost. Some of our bounds are a function of $\theta = \min_{1 \le k \le |A|} |I_k|$, the size of a least popular class. For simplicity, we use θ to denote both the class and its size. We prove a tight bound of m for the PoA, and an almost tight bound of $\theta + 1$, implying that the existence of a single small class guarantees low PoA. We start with the lower bound based on the number of servers. Specifically, we show that the PoA may be $m - \epsilon$ for any $\epsilon > 0$.

Theorem 3. *For any $m \ge 2$ servers and any $\epsilon > 0$, there exists an instance G for which $PoA(G) > m - \epsilon$.*

Proof. Let k be an integer such that $\frac{1}{m^k} \le \epsilon$. Consider an instance G with $n = m^{k+3}$ players, $U = n$ and a single class. Let f be the allocation in which all the players are allocated to a single server. The cost of each player in f is $c_1 = n + 1 = m^{k+3} + 1$. A player migrating to an empty server would have a cost of $1 + U = n + 1 = c_1$. Thus, f is stable. On the other hand, consider an allocation f' in which the players are equally distributed between the servers. Each server is allocated with m^{k+2} players, each having cost $c_1' = m^{k+2} + m$. Therefore,

$$PoA(G) \ge \frac{c_1}{c_1'} = \frac{m^{k+3} + 1}{m^{k+2} + m} > m - \frac{1}{m^k} \ge m - \epsilon.$$

$\qquad \square$

In order to prove the upper bound, we first provide several lower bounds on OPT. Let $d = \max(\frac{n}{m}, \sqrt{U})$.

Claim 4. $OPT \geq \max(\frac{n+\frac{U}{\theta}}{m}, \lceil \frac{n}{m} \rceil, \frac{U}{\theta}, 2\sqrt{U}, d + \frac{U}{d})$.

When $\theta \leq \frac{n}{m}$, we can bound OPT further as a function of θ and U.

Claim 5. *If* $\theta \leq \frac{n}{m}$, *then* $OPT \geq \theta + \frac{U}{\theta}$.

Theorem 6. *For any resource allocation game* G *with multiple resource classes,* $PoA(G) \leq m$.

Proof. Let f be a stable allocation, and let j_1 be a server such that $c_f(j_1) = c_{max}(f)$. Let i be a class with minimal group-size on j_1. Thus, $c_1 = L_1 + \frac{U}{L_{j_1,i}}$ is the maximal cost of a player in f. We show that $c_1 \leq n + \frac{U}{\theta}$. By Claim 4, this implies that the PoA is at most m.

If j_1 is the only server that services players from class i then $L_{j_1,i} \geq \theta$. Thus, $c_1 \leq n + \frac{U}{\theta}$.

If players from class i are assigned in f to more than a single server, let $j_2 \neq j_1$ be a least loaded server that services class-i players in f. Denote $\ell_1 = L_{j_1,i}$ and $\ell_2 = L_{j_2,i}$. The cost of a class-i player on j_2 is $c_2 = L_2 + \frac{U}{\ell_2}$. Since f is stable, a migration of an i-player from j_1 to j_2 is not beneficial. Combining the fact that $c_2 \leq c_1$, we get

$$L_2 + \frac{U}{\ell_2} \leq L_1 + \frac{U}{\ell_1} \leq L_2 + 1 + \frac{U}{\ell_2 + 1}. \tag{2}$$

Equation (2) implies that $U \leq \ell_2(\ell_2 + 1)$.

On the other hand, a migration of an i-player from j_2 to j_1 is also not beneficial. Thus, $L_2 + \frac{U}{\ell_2} \leq L_1 + 1 + \frac{U}{\ell_1+1}$ and we get

$$L_2 + 1 + \frac{U}{\ell_2 + 1} \leq L_2 + 1 + \frac{U}{\ell_2} \leq L_1 + 2 + \frac{U}{\ell_1 + 1}. \tag{3}$$

Combining Eqs. (2) and (3), we conclude that

$$U \leq \min(2\ell_1(\ell_1 + 1), \ell_2(\ell_2 + 1)). \tag{4}$$

If class-i players are allocated to exactly two servers, the analysis is technically involved and is omitted due to space constraints.

If class-i players are allocated to more than two servers then since j_2 is the least loaded server with class-i players, except possibly j_1, we have $\ell_2 \leq L_2 < \frac{n}{2}$ and $c_1 \leq L_2 + 1 + \frac{U}{\ell_2+1} \leq L_2 + 1 + \ell_2 < n$. Thus, for every possible allocation of class-i players, we showed that $c_1 \leq n + \frac{U}{\theta} \leq m \cdot OPT$. □

Our next bound depends on the size of the smallest class. We start with the upper bound.

Theorem 7. *For any resource allocation game* G *with multiple resource classes, and any* $\epsilon > 0$, $PoA(G) \leq \theta + 1$.

Proof. Let f be a stable allocation, and let j be a server such that $c_f(j) = c_{max}(f)$. Let L_1 be the load on j and let L_0 be the load on the least loaded server in f. If $L_1 \leq \lceil \frac{n}{m} \rceil$ then $c_{max}(f) \leq \lceil \frac{n}{m} \rceil + U$. Otherwise, by the pigeonhole principle, $L_0 < \lceil \frac{n}{m} \rceil$. Since f is stable, $c_f(j) \leq L_0 + U + 1 \leq \lceil \frac{n}{m} \rceil + U$. By Claim 4, $OPT \geq \max(\lceil \frac{n}{m} \rceil, \frac{U}{\theta})$. Thus, $PoA \leq \frac{\lceil \frac{n}{m} \rceil + U}{OPT} \leq \theta + 1$. $\qquad\square$

This bound is almost matched.

Theorem 8. *For any $\theta \geq 1$ and $\epsilon > 0$, there exists an instance G for which $PoA(G) > \theta - \epsilon$.*

Proof. Given ϵ and θ, let U be a constant such that $\epsilon \geq \frac{\theta^3}{U + \theta^2}$. Consider an instance with $n = U(1 - \frac{1}{\theta})$ players from two classes, where $|I_1| = \theta$ and $|I_2| = n - \theta$. Let $m = n/\theta$. Note that U can be selected such that n and m are integers.

Let f be the allocation in which all the players are allocated to a single server. Players of I_1 have the max-cost in f, which is $c_1 = n + \frac{U}{\theta}$. A player migrating to an empty server would have a cost of $U + 1$. Since $U = n + \frac{U}{\theta} = c_1$, such a migration is not beneficial. Thus, f is stable. On the other hand, consider an allocation f' in which the players are equally distributed between the servers, each server accommodating θ players from the same class. All the players have cost $c' = \theta + \frac{U}{\theta}$. Therefore,

$$PoA(G) \geq \frac{c_1}{c'} = \frac{n + \frac{U}{\theta}}{\theta + \frac{U}{\theta}} = \frac{U}{\theta + \frac{U}{\theta}} \geq \theta - \epsilon.$$

$\qquad\square$

4   Equilibrium Inefficiency - Price of Stability

In this section we analyze the Price of Stability with respect to the max-cost objective. For systems with arbitrary number of servers, m, we show that $2 - \frac{1}{m} \leq PoS \leq 2$. For two servers, the lower bound is tight. Specifically, we present an $O(|A| \log |A| + n)$-time algorithm for calculating a NE assignment that achieves max-cost at most $\frac{3}{2} OPT$. The algorithm is omitted from this extended abstract.

Our main result is an algorithm for arbitrary number of servers. The algorithm combines load-balancing ideas used in packing algorithms, such as element-grouping and handling of elements in decreasing-size order, together with ideas used in algorithmic game theory, such as performing BRD in a specific order. We begin with a lower bound of $2 - \frac{1}{m}$.

Theorem 9. *For every $\epsilon > 0$ and a system with $m \geq 2$ servers, there exists an instance G such that $PoS(G) > 2 - \frac{1}{m} - \epsilon$.*

Proof. Given $\epsilon > 0$, let $n = \max(\lceil \frac{4(m-1)}{\epsilon} \rceil, 4m)$. Consider an instance G with $m \geq 2$ servers, and $A = \{a_1, a_2\}$, where a single player belongs to class a_1 and all other players belong to class a_2. Let $U = \frac{n-1}{m-1} - 2$. A possible allocation for this instance is illustrated in Fig. 1(a). The players who belong to a_2 are split evenly among $m-1$ servers and the player of a_1 is solely allocated to the remaining server. The maximal cost for this allocation is for players who belong to a_2 and is $c_1 = \frac{n-1}{m-1} + 1 - \frac{2(m-1)}{n-1}$. The only NE (up to server renaming) for this instance is illustrated in Fig. 1(b). The player of a_1 has the maximal cost for this allocation $c_2 = \frac{n}{m} + \frac{n-1}{m-1} - 2$. A player of a_2 has cost at most $c_3 = \frac{n}{m} + \frac{U}{\frac{n}{m}-1}$, a player of a_2 migrating to a different server would have cost at least $c_4 = \frac{n}{m} + 1 + \frac{U}{\frac{n}{m}+1}$. Since $n \geq 4m$ and $m \geq 2$, $\frac{n-1}{m-1} - 1 < \frac{n}{m}$ and $U < \frac{n}{m} - 1$. Thus, $c_3 < c_4$ and the allocation is stable. We conclude that the PoS is at least

$$\frac{c_2}{c_1} = \frac{\frac{n}{m} + \frac{n-1}{m-1} - 2}{\frac{n-1}{m-1} + 1 - \frac{2(m-1)}{n-1}} \geq \frac{\frac{n}{m} + \frac{n-1}{m-1} - 2}{\frac{n-1}{m-1} + 1} \geq 2 - \frac{1}{m} - \frac{4(m-1)}{n} \geq 2 - \frac{1}{m} - \epsilon.$$

$\square$

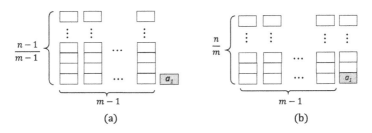

Fig. 1. (a) An optimal non-stable allocation, (b) A best NE.

4.1 An Algorithm for Multiple Servers

For a system with an arbitrary number of servers, we present a polynomial time algorithm that constructs a NE with max-cost at most $2OPT$. We use the term *big classes* when referring to classes with at least $\frac{n}{m}$ players. Similar to the case $m = 2$, Algorithm 1, given below, assigns complete classes to servers while only splitting big classes. This initial assignment is similar to Longest Processing Time (LPT) algorithm for job scheduling [6], that is, it assigns the sets greedily, in non-increasing order, on a least loaded server. If the resulting assignment is not stable, a stabilization phase is performed. This phase consists of migrations of complete classes or sequences of supervised improving steps. The improvement steps are in 'Follow-a-leader' phases. That is, once one member of a class performs a beneficial migration, an *identical* migration is considered for other members of his class. While it is complex to analyze the change in the social cost of arbitrary sequence of improving steps, we are able to analyze it for this structured stabilization phase. Recall that $d = \max(\frac{n}{m}, \sqrt{U})$.

Algorithm 1. An algorithm for finding a NE achieving max-cost at most $2OPT$.

Let $d = \max(\sqrt{U}, \frac{n}{m})$.

1. Consider the players according to their classes.
2. Partition any class I_k such that $I_k \geq d$ to $\left\lfloor \frac{I_k}{d} \right\rfloor$ sets of equal sizes (up to a rounding difference of 1).
3. Sort the resulting sets by their size in decreasing order.
4. Consider the sets according to the sorted order, assign all the players of the next set to a least loaded server.
5. If the schedule is not stable, perform a Stabilization Phase (Algorithm 2).

Let f denote the allocation produced in step 4. We start by characterizing f and show that $c_{max}(f) < 2OPT$. We then consider the case that f is not stable and the stabilization phase is applied. We show that this phase is guaranteed to converge to a NE allocation f' for which $c_{max}(f') < 2OPT$. We first characterize some cases in which any NE f_0 fulfills $c_{max}(f_0) < 2OPT$, and then analyze the stabilization phase for the remaining cases.

Claim 10. *The maximal load on a server in the allocation f is at most $2d - 1$.*

Proof. Assume by contradiction that there is a server s with load at least $2d$. Step 2 guarantees that the maximal set-size is at most $2d - 1$. Thus, there are at least two different sets allocated to s. Let Γ be the first set allocated to s that increases the load beyond $2d - 1$. Let ℓ be the load on s before Γ is added. Since the sets are ordered by decreasing order of their sizes, $\Gamma \leq \ell$. If $\ell \geq \frac{n}{m}$ then by the pigeonhole principle there is a server s_0 such that $L_{s_0} < \frac{n}{m}$, contradicting the assignment of Γ to s. If $\ell < \frac{n}{m}$ then $|\Gamma| + \ell \leq 2\ell < \frac{2n}{m} \leq 2d$, contradicting the assumption that s gets load at least $2d$. $\qquad \Box$

Lemma 11. $c_{max}(f) < 2OPT$.

Proof. Consider a server s such that $c_{max}(f) = c_f(s)$. By Claim 10 the maximal load on s is at most $2d - 1$. If all the players in s belong to the same class, $c_f(s) \leq 2d-1+\frac{U}{d} < 2d+\frac{U}{d}$. By Claim 4, $OPT \geq d+\frac{U}{d}$. Thus, $c_{max}(f) < 2OPT$. Let θ_0 be the last set assigned to s, if s is assigned with players of different classes, then $\theta_0 < \frac{n}{m}$ since the sets are assigned by LPT order. By the pigeonhole principal, the load on s is at most $\frac{n}{m} + \theta_0$. Thus, $c_f(s) \leq \frac{n}{m} + \theta_0 + \frac{U}{\theta_0}$. Since $\theta \leq \theta_0 \leq \frac{n}{m}$ and $x + \frac{U}{x}$ is a convex function, using Claims 4 and 5, we conclude $\theta_0 + \frac{U}{\theta_0} \leq \max(\theta + \frac{U}{\theta}, \frac{n}{m} + \frac{Um}{n}) \leq OPT$ and $c_f(s) < 2OPT$. $\qquad \Box$

Next, we show the stabilization phase converges to a stable assignment. The proof of the following claim is based on analyzing the change in the potential function $\Phi(f)$ defined in (1). We show that every iteration of Step 1 of Algorithm 2 reduces the potential. By Theorem 1, this is valid also for Step 2.

Algorithm 2. Stabilization Phase

Repeat until convergence:

1. While there exists a server s_1 and a class I_k such that all players from I_k are on s_1 and $L_1 \geq |I_k| + \frac{n}{m}$, move I_k from s_1 to some server s_2 for which $L_2 < \frac{n}{m}$.
2. Perform a 'follow a leader' sequence of improving steps:
    2.1. Let i_1 be some player that has a beneficial move. Assume $i_1 \in I_k$ and denote by s_1 the server to which i_1 is assigned.
    2.2. Let i_1 perform a beneficial step from s_1 to some server s_2.
    2.3. As long as there exists another unsatisfied player $i \in I_k$ assigned to s_1, for which migrating to s_2 is beneficial, let i migrate to s_2.

Claim 12. *The stabilization phase converges to a NE.*

We turn to analyze the cost of the stable assignment f' produced by the stabilization phase. For some cases, a 2-ratio can be shown for any stable assignment.

Lemma 13. *If $U \leq \frac{n}{m}$ or $\frac{n}{m} < U < 4$ or $\theta = 1$, then for any NE f' it holds that $c_{max}(f') \leq 2OPT$.*

For the remaining cases, we analyze the outcome of the stabilization phase. We use below known properties of assignment produced by LPT algorithm.

Claim 14. *If f is not stable then $U < 2d$.*

Proof. By Claim 10, the maximal load on a server in f is at most $2d-1$. Let i be a player in server s_1 with a beneficial move to s_2. The load difference between s_1 and s_2 is at most $2d-1$. The big classes are equally distributed in Step 2 to sets of size at least d. Since $d \geq \frac{n}{m}$ and the sets are allocated in non-increasing order of size, servers with a set of a big class are only assigned players of that class. Thus, since $d \geq \sqrt{U}$, players of big classes can only have a beneficial move to servers not servicing the same class. Players of small classes are all in the same set generated in Step 1 and are all allocated to the same server. Obviously, such players can only have a beneficial move to a server not assigned with their class. Let Γ be the last set assigned to s_1 in Step 4. Since the sets are assigned in non-increasing order of size, $L_1 - L_2 < |\Gamma|$ and the cost of i prior to the improving step is at most $c_1 = L_1 + \frac{U}{|\Gamma|}$. The cost after the step is $c_2 = L_2 + 1 + U$. Since $c_2 < c_1$ we have $L_2 + 1 + U < L_1 + \frac{U}{|\Gamma|}$. Thus, $U(\frac{|\Gamma|-1}{|\Gamma|}) < L_1 - L_2 - 1 \leq |\Gamma| - 1$ and $U \leq |\Gamma| \leq 2d - 1$. □

Lemma 15. *If $U \geq 4$ and $\theta > 1$, then Step 2 of the stabilization phase results in an allocation with at least two players in any class allocated to a server.*

Lemma 16. *The maximal load on a server in the allocation f' is at most $2d-1$.*

We summarize with the following Theorem.

Theorem 17. *Algorithm 1 produces a NE assignment with max-cost at most* $2OPT$.

Proof. If the allocation f generated in Step 3 is stable then by Lemma 11 its max-cost is at most $2OPT$. If f is not stable, and $\theta = 1$ or $U \leq \frac{n}{m}$ or $\frac{n}{m} < U < 4$, then by Lemma 13, any NE has max-cost at most $2OPT$. If f is not stable, $\theta > 1$, $U > \frac{n}{m}$ and $U \geq 4$ then by Claim 12 and Lemma 15, the stabilization phase converges to a stable allocation f' in which the smallest set on each server is of size at least 2. Assume by contradiction that $c_{max}(f') > 2OPT$. Let s be a server such that $c_{f'}(s) > 2OPT$. The cost of s is at most $L_{f'}(s) + \frac{U}{2}$. Using Claim 14 we have $U < 2d$ thus $c_{f'}(s) < L_{f'}(s) + d$ and $L_{f'}(s) > d$. If there is a single class allocated to s then $c_{f'}(s) \leq L_{f'}(s) + \frac{2d}{L_{f'}(s)}$. By Lemma 16, $L_{f'}(s) < 2d$ and $c_{f'}(s) < 2d$. If there are multiple classes allocated to s then by Lemma 15 the smallest set of a players Γ who belong to the same class on s is at least 2. Since Γ was not moved by Step (1) of the stabilization phase, we conclude $c_{f'}(s) \leq \frac{n}{m} - 1 + |\Gamma| + \frac{U}{|\Gamma|} \leq d - 1 + |\Gamma| + \frac{U}{|\Gamma|}$. Since $2 \leq |\Gamma| \leq \frac{n}{m}$ we have $|\Gamma| + \frac{U}{\Gamma} \leq \max(2 + \frac{U}{2}, \frac{n}{m} + \frac{Um}{n} \leq d + \frac{U}{d})$. Claim 4 implies that $2OPT \geq 2d + \frac{2U}{d}$ and also $2d + \frac{U}{2d} \geq d - 1 + |\Gamma| + \frac{U}{|\Gamma|}$. Finally, since $2d + \frac{2U}{d} \geq 2d + \frac{U}{2d}$, we get $c_{max}f'(s) \leq 2OPT$. □

5   Conclusions and Open Problems

We studied a resource-allocation game with multiple resource classes in which user's cost function encompasses both negative and positive, class-dependent, congestion effects. Our study of the game reveals that even for the basic model of unit-load players and identical servers, the equilibrium inefficiency may by very high. On the other hand, an assignment whose cost is at most twice the optimum exists and can be calculated in poly-time. We list below some open problems and possible directions for future work.

1. Heterogeneous systems: our work considers systems with identical servers and unit-load requirements. One possible generalization is to study systems with unrelated servers and/or non-identical load requirements. In the classic load balancing game, there is a significant difference between the results regarding related and unrelated systems. It would be interesting to study the corresponding differences in the multi-class model.
2. Players with class preferences or with multiple classes: In our work players belong to a single class. In a possible generalization of this game (studied in [15] for the centralized model), a player may belong to several classes and has preferences regarding his class. This scenario fits for example MoD systems in which a client is ready to see one of several movies, and provides his preferences for broadcast. In the corresponding game, the utility of a player depends also on the class to which it is assigned. Another direction is to study systems in which a player requires more than a single resource for his processing. Thus, a player may belong to multiple classes and needs to pay his share in the activation cost of all the resources he needs.

3. We calculated inefficiency with respect to the max-cost objective function. Future work could also consider other objective functions such as sum-cost.
4. BRD convergence time: We have shown that BRD converges within an upper bound of $O(n^4)$ steps. A lower bound of $\Omega(n \log n)$ steps follows from the analysis in [4] for a single class. Closing the gap and providing a tight bound for BRD convergence time remains open.
5. Strong Equilibrium: In a work in progress we have shown that a SE may not exist for $U > 2$, while for $U = 0$ a SE always exist. The existence of SE for $0 < U \leq 2$ is an open question. Characterizing conditions in which an SE exists and analyzing SE inefficiency are additional open directions.
6. Capacitated Model: We assumed that servers have unlimited capacity. Studying the capacitated game, in which servers have limited storage and/or limited load capacities arise new challenges.

References

1. Anshelevich, E., Dasgupta, A., Kleinberg, J.M., Tardos, É., Wexler, T., Roughgarden, T.: The price of stability for network design with fair cost allocation. SIAM J. Comput. **38**(4), 1602–1623 (2008)
2. Anshelevich, E., Dasgupta, A., Kleinberg, J.M., Tardos, É., Wexler, T., Roughgarden, T.: The price of stability for network design with fair cost allocation. In: Symposium on the Foundations of Computer Science (FOCS), pp. 295–304 (2004)
3. Chen, B., Gürel, S.: Efficiency analysis of load balancing games with and without activation costs. J. Sched. **15**(2), 157–164 (2012)
4. Feldman, M., Tamir, T.: Conflicting congestion effects in resource allocation games. J. Oper. Res. **60**(3), 529–540 (2012)
5. Fang, X., Zhe, X., Yuzhong, Z., Qingguo, B.: Scheduling games on uniform machines with activation cost. Theo. Comput. Sci. **580**, 28–35 (2015)
6. Graham, R.: Bounds on multiprocessing timing anomalies. SIAM J. Appl. Math. **17**, 263–269 (1969)
7. Koutsoupias, E., Papadimitriou, C.: Worst-case equilibria. Comput. Sci. Rev. **3**(2), 65–69 (2009)
8. Lin, L., Yan, Y., He, X., Tan, Z.: The PoA of scheduling game with machine activation costs. In: Chen, J., Hopcroft, J.E., Wang, J. (eds.) FAW 2014. LNCS, vol. 8497, pp. 182–193. Springer, Heidelberg (2014). doi:10.1007/978-3-319-08016-1_17
9. Monderer, D., Shapley, L.S.: Potential games. Game. Econ. Behav. **14**, 124–143 (1996)
10. Prodan, R., Ostermann, S.: A survey and taxonomy of infrastructure as a service and web hosting cloud providers. In: IEEE/ACM International Conference on Grid Computing, pp. 17–25 (2009)
11. Rosenthal, R.W.: A class of games possessing pure-strategy Nash equilibria. Int. J. Game Theory **2**, 65–67 (1973)
12. Roughgarden, T.: Chapter 18: Routing games. In: Nisan, N., Roughgarden, T., Tardos, E., Vazirani, V.V. (eds.) Algorithmic Game Theory. Cambridge University Press, Cambridge (2007)
13. Shachnai, H., Tamir, T.: On two class-constrained versions of the multiple knapsack problem. Algorithmica **29**, 442–467 (2001)

14. Shachnai, H., Tamir, T.: Tight bounds for online class-constrained packing. Theoret. Comput. Sci. **321**(1), 103–123 (2004)
15. Tamir, T., Vaksendiser, B.: Algorithms for storage allocation based on client preferences. J. Comb. Optim. **19**, 304–324 (2010)
16. Vöcking, B.: Chapter 20: Selfish load balancing. In: Nisan, N., Roughgarden, T., Tardos, T., Vazirani, V.V. (eds.) Algorithmic Game Theory. Cambridge University Press, Cambridge (2007)
17. Wolf, J.L., Yu, P.S., Shachnai, H.: Disk load balancing for video-on-demand systems. ACM Multimedia Syst. J. **5**, 358–370 (1997)

Tight Approximation Bounds for the Seminar Assignment Problem

Amotz Bar-Noy and George Rabanca$^{(\boxtimes)}$

Department of Computer Science, Graduate Center, CUNY, New York City, USA
amotz@sci.brooklyn.cuny.edu, grabanca@gradcenter.cuny.edu

Abstract. The seminar assignment problem is a variant of the generalized assignment problem in which items have unit size and the amount of space allowed in each bin is restricted to an arbitrary set of values. The problem has been shown to be NP-complete and to not admit a PTAS. However, the only constant factor approximation algorithm known to date is randomized and it is not guaranteed to always produce a feasible solution.

In this paper we show that a natural greedy algorithm outputs a solution with value within a factor of $(1 - e^{-1})$ of the optimal, and that unless $NP \subseteq DTIME(n^{\log \log n})$, this is the best approximation guarantee achievable by any polynomial time algorithm.

Keywords: General assignment · Budgeted maximum coverage · Seminar assignment problem

1 Introduction

In the SEMINAR ASSIGNMENT problem (SAP) introduced in [8] one is given a set of seminars (or bins) B, a set of students (or items) I, and for each seminar b a set of integers K_b specifying the allowable number of students that can be assigned to the seminar. Unless otherwise specified, we assume that $0 \in K_b$ for any $b \in B$. For each student i and seminar $b \in B$ let $p(i, b) \in \mathbb{R}$ represent the profit generated from assigning student i to seminar b. A *seminar assignment* is a function $\mathcal{A} : J \to B$ where $J \subseteq I$ and we say that the assignment is feasible if $|\mathcal{A}^{-1}(b)| \in K_b$ for all $b \in B$, where $\mathcal{A}^{-1}$ is the pre-image of $\mathcal{A}$. The goal is to find a feasible seminar assignment $\mathcal{A}$ that maximizes the total profit:

$$p(\mathcal{A}) = \sum_{i \in J} p(i, \mathcal{A}(i)).$$

The problem has been introduced in [8] in a slightly less general version. In the original version, for each $b \in B$ the set K_b equals to $\{0\} \cup \{l_b, ..., u_b\}$ for some

---

This work is supported in part by grants from NSF CNS 1302563, by Navy N00014-16-1-2151, by NSF CNS 1035736, by NSF CNS 1219064. Any opinions, findings, and conclusions or recommendations expressed here are those of the authors and do not necessarily reflect the views of sponsors.

© Springer International Publishing AG 2017
K. Jansen and M. Mastrolilli (Eds.): WAOA 2016, LNCS 10138, pp. 170–182, 2017.
DOI: 10.1007/978-3-319-51741-4_14

lower and upper bounds $l_b, u_b \in \mathbb{N}$. The more general setting considered in this paper can be useful for example when a seminar doesn't just require a minimum number of students and has a fixed capacity, but in addition requires students to work in pairs and therefore would allow only an even number of students to be registered. In addition, this generalization also simplifies notation.

SAP is a variant of the classic GENERAL ASSIGNMENT problem (GAP) in which one is given m bins with capacity $B_1, ..., B_m$ and n items. Each item i has size $s(i, b)$ in bin b and yields profit $p(i, b)$. The goal is to find a packing of the items into the bins that maximizes total profit, subject to the constraint that no bin is overfilled. A GAP instance with a single bin is equivalent to the knapsack problem, and a GAP instance with unit profit can be interpreted as a decision version of the bin packing problem: can all items be packed in the m bins?

SAP is also related to the MAXIMUM COVERAGE problem (MC). In the classic version of the MC problem one is given a collection of sets $\mathcal{S} = \{S_1, ..., S_m\}$ and a budget B. The goal is to select a subcollection $\mathcal{S}' \subseteq \mathcal{S}$ with cardinality less than or equal to B such that $|\cup_{S \in \mathcal{S}'} S|$ is maximized.

The algorithms with the best approximation ratio for both MC and GAP are greedy algorithms and the approximation bounds have been proved with similar techniques. In this paper we show how to extend these analysis techniques to SAP.

Related Work. In [8] the authors show that SAP is NP-complete even when $K_b = \{0, 3\}$ for all $b \in B$ and $p(i, b) \in \{0, 1\}$ for any $i \in I$. Moreover, they show that SAP does not admit a PTAS by providing a gap-preserving reduction from the 3-bounded 3-dimensional matching problem. In [1] the authors investigate the approximability of the problem and provide a randomized algorithm which they claim outputs a solution that in expectation has value at least $1/3.93$ of the optimal. In [2] this result is revised and the authors show that for any $c \geq 2$, their randomized algorithm outputs a feasible solution with probability at least $1 - \min\{\frac{1}{c}, \frac{e^{c-1}}{c^c}\}$ and has an approximation ratio of $\frac{e-1}{(2c-1)\cdot e}$.

The GAP is well studied in the literature, with [3,9] surveying the existing algorithms and heuristics for multiple variations of the problem. In [11] the authors provide a 2-approximation algorithm for the problem and in [4] it is shown that any α-approximation algorithm to the knapsack problem can be transformed into a $(1+\alpha)$-approximation algorithm for GAP. In [6] tight bounds for the GAP are given showing that no polynomial time algorithm can guarantee a solution within a factor better than $(1 - e^{-1})$, unless $P = NP$, and providing an LP-based approximation which for any $\epsilon > 0$ outputs a solution with profit within a $(1 - e^{-1} - \epsilon)$ factor of the optimal solution value.

The GAP with minimum quantities, in which a bin cannot be used if it is not packed at least above a certain threshold, is introduced in [8]. Because items have arbitrary size, it is easy to see that when a single bin is given and the lower bound threshold equals the bin capacity, finding a feasible solution with profit greater than zero is equivalent to solving SUBSET SUM. Therefore, in its most general case the problem cannot be approximated in polynomial time, unless $P = NP$.

In [5,10] the authors study the problem of maximizing a non-decreasing submodular function f satisfying $f(\emptyset) = 0$ under a cardinality constraint. They show that a simple greedy algorithm achieves an approximation factor of $(1-e^{-1})$ which is the best possible under standard assumptions. Vohra and Hall note that the classic version of the maximum coverage problem belongs to this class of problems [13]. When each set S_i in the MC problem is associated with a cost $c(S_i)$ the BUDGETED MAXIMUM COVERAGE problem asks to find a collection of sets $\mathcal{S}'$ covering the maximum number of elements under the (knapsack) constraint that $\sum_{S_i \in \mathcal{S}'} c(S_i) \leq B$ for some budget $B \in \mathbb{R}$. In [7] the authors show that the greedy algorithm combined with a partial enumeration of all solutions with small cardinality also achieves a $(1 - e^{-1})$ approximation guarantee, and provide matching lower bounds which hold even in the setting of the classic MC problem (when all sets have unit cost). In [12] Sviridenko generalizes the algorithm and proof technique to show that maximizing any monotone submodular function under a knapsack constraint can be approximated within $(1 - e^{-1})$ as well.

Contributions. In Sect. 2, by a reduction from the MAXIMUM COVERAGE problem, we show that there exists no polynomial time algorithm that guarantees an approximation factor larger than $(1 - e^{-1})$, unless $NP \subseteq DTIME(n^{\log \log n})$. In Sect. 4 we present a greedy algorithm that outputs a solution that has profit at least $\frac{1}{2} \cdot (1 - e^{-1})$ of the optimal solution. The algorithm is based on the observation that when the required number of students in each seminar is fixed, the problem is solvable in polynomial time. Finally, in Sect. 5 we show how this algorithm can be improved to guarantee an approximation bound of $(1 - e^{-1})$.

2 Hardness of Approximation

In this section we show that the problem is hard to approximate within a factor of $(1 - e^{-1} + \epsilon)$, $\forall \epsilon > 0$, even for the case when for each $b \in B$ the set K_b equals $\{0, n\}$ for some integer n, and the profit for assigning any student to any seminar is either 0 or 1. We prove this result by showing that such restricted instances of SAP are as hard to approximate as the MAXIMUM COVERAGE problem defined below.

Definition 1. *Given a collection of sets $\mathcal{S} = \{S_1, ..., S_m\}$ and an integer k, the* MAXIMUM COVERAGE *(MC) problem is to find a collection of sets $\mathcal{S}' \subseteq \mathcal{S}$ such that $|\mathcal{S}'| \leq k$ and the union of the sets in $\mathcal{S}'$ is maximized.*

In [7] it is shown that the MC problem is hard to approximate within a factor of $(1 - e^{-1} + \epsilon)$, unless $NP \subseteq DTIME(n^{\log \log n})$. We use this result to prove the following:

Theorem 1. *For any $\epsilon > 0$ the SAP is hard to approximate within a factor of $(1 - e^{-1} + \epsilon)$ unless $NP \subseteq DTIME(n^{\log \log n})$.*

Proof. To prove the theorem we create a SAP instance for any given MC instance and show that from any solution of the SAP instance we can create a solution for the MC instance with at least equal value, and that the optimal solution of the SAP instance has value at least equal to the optimal solution of the MC instance. Therefore, an α-approximation algorithm for SAP can be transformed into an α-approximation algorithm for MC.

Given a MC instance, let $U = \cup_{S \in \mathcal{S}} S$ and $n = |U|$. For each set $S \in \mathcal{S}$ let b_S be a seminar with the allowable number of students $K_b = \{0, n\}$, and for each element $e \in U$ let i_e be a student in I. The profit of a student i_e assigned to a seminar b_S is 1 if the element e belongs to the set S and 0 otherwise. In addition, let $d_1, ..., d_{n*(k-1)}$ be dummy students that have profit 0 for any seminar.

We first show that any feasible assignment $\mathcal{A}$ corresponds to a valid solution to the given MC instance. Since every seminar requires exactly n students and there are exactly $k \cdot n$ students available, clearly at most k seminars can be assigned students in any feasible assignment. Let $\mathcal{S}' = \{S \in \mathcal{S} : \mathcal{A}(b_S) > 0\}$. It is easy to see that the number of elements in $\cup_{S \in \mathcal{S}'} S$ is at least equal to the profit $p(\mathcal{A})$ since a student i_e has profit 1 for a seminar b_S only if the set S covers element e.

It remains to show that for any solution to the MC instance there exists a solution to the corresponding SAP instance with the same value. Fix a collection of sets $\mathcal{S}' \subseteq \mathcal{S}$ with $|\mathcal{S}'| \leq k$. For every $e \in \cup_{S \in \mathcal{S}'} S$ let S_e be a set in $\mathcal{S}'$ that contains e and let $\mathcal{A}(i_e) = b_{S_e}$. Then, assign additional dummy students to any seminar with at least one student to reach the required n students per seminar. Clearly, the profit of the assignment $\mathcal{A}$ is equal to the number of elements covered by the collection $\mathcal{S}'$, which proves the theorem. $\square$

3   Seminars of Fixed Size

In this section we show that when the allowable number of students that can be assigned to any seminar b is a set $K = \{0, k_b\}$ for some integer k_b, SAP can be approximated within a factor of $(1 - e^{-1})$ in polynomial time. This introduces some of the techniques used in the general case in a simpler setting.

For an instance of the SAP, a *seminar selection* is a function $S : B \to \mathbb{N}$ with the property that $S(b) \in K_b$ for any $b \in B$. We say that S is feasible if $\sum_{b \in B} S(b) \leq |I|$. In other words, a seminar selection is a function that maps each seminar to the number of students to be assigned to it. A seminar selection S *corresponds* to an assignment $\mathcal{A}$ if for any seminar b the number of students assigned by $\mathcal{A}$ to b is $S(b)$. We slightly abuse notations and denote by $p(S)$ the maximum profit over all seminar assignments corresponding to the seminar selection S; we call $p(S)$ the profit of S. In the remainder of this paper for a graph $G = (V, E)$ we denote the subgraph induced by the vertices of $X \subseteq V$ by $G[X]$.

Definition 2. *Given a SAP instance let* $V_b = \{v_{b,1}, ..., v_{b,k_b}\}$ *for every* $b \in B$ *and let* $V = \cup_{b \in B} V_b$. *The* **bipartite representation** *of the instance is the complete* *bipartite graph* $G = (V \cup I, E)$ *with edge weights* $w(v_b, i) = p(i, b)$ *for every* $v_b \in V_b$. *The* bipartite representation *of a seminar selection* S *is the graph* $G[V_S \cup I]$ *where* $V_S = \cup_{b \in B} V_{S,b}$ *and* $V_{S,b} = \{v_{b,1}, ..., v_{b,S(b)}\}$ *for every* $b \in B$.

Lemma 1. *For any SAP instance and any feasible seminar selection* S, $p(S)$ *is equal to the value of the maximum weight matching in the bipartite representation of* S.

Proof. Let $G_S = (V_S \cup I, E)$ be the bipartite representation of S. First observe that any matching M of G_S that matches all the vertices of V_S can be interpreted as an assignment $\mathcal{A}_M$ of equal value by setting $\mathcal{A}_M(i) = b$ whenever vertex $i \in I$ is matched by M to a vertex in $V_{S,b}$. Since G_S is complete and has non-negative edge weights, there exists a maximum weight matching that matches all the vertices of V_S.

Similarly, any feasible assignment for the SAP instance can be interpreted as a matching $M_\mathcal{A}$ of equal value, which proves the lemma.   □

Definition 3. *For a given finite set* A, *a set function* $f : 2^A \to \mathbb{R}$ *is submodular if for any* $X, Y \subseteq A$ *it holds that:*

$$f(X) + f(Y) \geq f(X \cup Y) + f(X \cap Y).$$

Sviridenko shows that certain submodular functions can be maximized under knapsack constraints, which will be useful in proving Theorem 3:

Theorem 2 ([12]). *Given a finite set* A, *a submodular, non-decreasing, non-negative, polynomially computable function* $f : 2^A \to \mathbb{R}$, *a budget* $L \geq 0$, *and costs* $c_a \geq 0$, $\forall a \in A$, *the following optimization problem is approximable within a factor of* $(1 - e^{-1})$ *in polynomial time:*

$$\max_{X \subseteq A} \left\{ f(X) : \sum_{x \in X} c_x \leq L \right\}$$

We relate now the value of a maximum weight matching in a bipartite graph to the notion of submodularity.

Definition 4. *For an edge weighted bipartite graph* $G = (A \cup B, E)$, *the* partial maximum weight matching function $f : 2^A \to \mathbb{R}$ *maps any set* $S \subseteq A$ *to the value of the maximum weight matching in* $G[S \cup B]$.

Lemma 2. *Let* f *be the partial maximum weight matching function for a bipartite graph* $G = (A \cup B, E)$ *with non negative edge weights. Then* f *is submodular.*

Proof. Fix two sets $X, Y \subseteq A$ and let $M_\cap$ and $M_\cup$ be two matchings for the graphs $G[(X \cap Y) \cup B]$ and $G[(X \cup Y) \cup B]$ respectively. To prove the lemma it is enough to show that it is possible to partition the edges in $M_\cap$ and $M_\cup$ into two disjoint matchings M_X and M_Y for the graphs $G[X \cup B]$ and $G[Y \cup B]$ respectively.

The edges of $M_\cap$ and $M_\cup$ form a collection of alternating paths and cycles. Let $\mathcal{C}$ denote this collection and observe that no cycle of $\mathcal{C}$ contains vertices from $X \setminus Y$ or $Y \setminus X$. This holds because $M_\cap$ does not match those vertices.

Let $\mathcal{P}_X$ be the set of paths in $\mathcal{C}$ with at least one vertex in $X \setminus Y$ and let $\mathcal{P}_Y$ be the set of paths in $\mathcal{C}$ with at least one vertex in $Y \setminus X$. Two such paths are depicted in Fig. 1.

Claim 1. $\mathcal{P}_X \cap \mathcal{P}_Y = \emptyset$.

Proof of claim: Assume by contradiction that there exists a path $P \in \mathcal{P}_X \cap \mathcal{P}_Y$. Let x be a vertex in $X \setminus Y$ on path P and similarly let y be a vertex in $Y \setminus X$ on path P. Observe that since neither x nor y belong to $X \cap Y$ they do not belong to the matching $M_\cap$ by definition, and therefore they are the endpoints of the path P. Moreover, since both x and y are in A, the path P has even length and since it is an alternating path, either the first or last edge belongs to $M_\cap$. Therefore $M_\cap$ matches either x or y contradicting its definition.   □

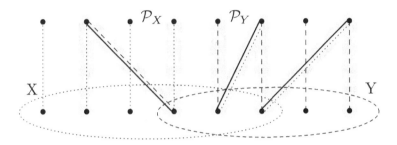

Fig. 1. $M_{X \cup Y}$ matches each vertex in $X \cup Y$ to the vertex directly above it. $M_{X \cap Y}$ is depicted with contiguous segments, M_X with dotted segments and M_Y with dashed segments. Two alternating paths of $\mathcal{P}$ are shown in light gray.

For a set of paths $\mathcal{P}$ we let $E(\mathcal{P}) = \{e \in P : P \in \mathcal{P}\}$. Moreover, let

$$M_X = (E(\mathcal{P}_X) \cap M_\cup) \cup (E(\mathcal{C} \setminus \mathcal{P}_X) \cap M_\cap)$$

and

$$M_Y = (E(\mathcal{P}_X) \cap M_\cap) \cup (E(\mathcal{C} \setminus \mathcal{P}_X) \cap M_\cup).$$

It is clear that $M_X \cup M_Y = M_\cap \cup M_\cup$ and $M_X \cap M_Y = M_\cap \cap M_\cup$. To prove the theorem it remains to show that M_X and M_Y are valid matchings for $G[X \cup B]$ and $G[Y \cup B]$ respectively. To see that M_X is a valid matchings for $G[X \cup B]$ observe first that no vertex of $Y \setminus X$ is matched by M_X since $\mathcal{P}_X$ does not intersect $Y \setminus X$ by Claim 1, and $M_\cap$ does not intersect $Y \setminus X$ by definition. Therefore, M_X only uses vertices of $X \cup B$. Second observe that every vertex $x \in X$ is matched by at most one edge of M_X since otherwise x belongs to either two edges of $M_\cup$ or two edges of $M_\cap$, contradicting the definition. This proves that M_X is a valid matching for $G[X \cup B]$; showing that M_Y is a valid matchings for $G[Y \cup B]$ is similar.   □

Theorem 3. *Any instance of SAP in which $|K_b| \leq 2$ for all $b \in B$ can be approximated in polynomial time to a factor of $(1 - e^{-1})$.*

Proof. Fix a SAP instance and for any $X \subseteq B$ let S_X be the seminar selection which allocates k_b students to any seminar in S and 0 students to any seminar in $B \setminus S$. Moreover, let G be the bipartite representation of the SAP instance and f be the partial maximum weight matching function for graph G. Denote by $G[V_X \cup I]$ the bipartite representation of S_X and let $g(X) = f(V_X)$. Since f is submodular by Lemma 2, it is easy to see that g is submodular as well. Assume by contradiction that there exist sets $X, Y \subseteq B$ such that the submodularity condition for g doesn't hold:

$$g(X) + g(Y) < g(X \cup Y) + g(X \cap Y). \tag{1}$$

Therefore, by definition of g we have

$$f(V_X) + f(V_Y) < f(V_X \cup V_Y) + g(V_X \cap V_Y),$$

contradicting the submodularity of f proven in Lemma 2.

Clearly g is also monotone, non-negative and polynomially computable. Let $c_b = k_b$, $\forall b \in B$, let $L = |I|$, and observe that S_X is feasible if and only if $\sum_{x \in X} c_x \leq L$. Moreover, by Lemma 1 and the definition of g, $g(X) = p(S_X)$ whenever the seminar selection S_X is feasible and therefore the proof follows from Theorem 2. □

4   A Constant Factor Greedy Algorithm

The algorithm presented in this section sequentially increments the number of students allocated to each seminar in a greedy fashion. It is similar in nature to the greedy algorithm of [7,12] but the details of the approximation guarantee proof are different. In the rest of this section we denote by $\mathcal{A}_S$ an optimal assignment for the seminar selection S. Remember that Lemma 1 shows that given feasible seminar selection S, an optimal seminar assignment $\mathcal{A}_S$ can be found in polynomial time.

We say that a seminar selection T is greater than a selection S (denoted by $T \succ S$) if $T(b) \geq S(b)$, $\forall b \in B$, and there exists $b \in B$ s.t. $T(b) > S(b)$. The cost of a seminar selection S is denoted by $c(S)$ and equals $\sum_{b \in B} S(b)$. When $T \succ S$ we define the *marginal cost* of T relative to S as the difference between the cost of T and the cost of S:

$$c_S(T) = c(T) - c(S)$$

Similarly, we define $p_S(T) = p(T) - p(S)$, the *marginal profit* of T relative to S. We say that T is an *incrementing selection* for a seminar selection S if $T \succ S$ and there exists a single seminar for which the selection T allocates more students than selection S; more precisely, the cardinality of the set $\{b \in B : T(b) > S(b)\}$ is 1. For a selection S we denote the set of incrementing seminar selections that are feasible by $inc(S)$.

We are now ready to present our algorithm:

Greedy

1. $S_0 = $ initial seminar selection;
2. $i = 0$;
3. While $inc(S_i) \neq \emptyset$:
   (a) $S_{i+1} \leftarrow \arg\max_{S' \in inc(S_i)} (p(S') - p(S_i))/(c(S') - c(S_i))$;
   (b) $i \leftarrow i + 1$
4. $\mathcal{A}_1 \leftarrow \mathcal{A}_{S_i}$;
5. $\mathcal{A}_2 \leftarrow$ maximum assignment to any single seminar b for which $S_0(b) = 0$;
6. Return $\max \mathcal{A}_1, \mathcal{A}_2$;

In this section we analyze the algorithm starting from an empty initial seminar selection. In the following section we show that by running the algorithm repeatedly with different initial seminar selections, the approximation guarantee can be improved.

Observe that the cardinality of $inc(S)$ is never greater than $|B| \cdot |I|$ and is therefore polynomial in the size of the input. Thus, using the maximum weight matching reduction from the proof of Lemma 1, step 3(a) of the algorithm can be performed efficiently.

Definition 5. *For a seminar selection S and a tuple (b, k_b) with $b \in B$ and $k_b \in \mathbb{N}$, let $S \oplus (b, k_b)$ denote the seminar selection S' with $S'(b) = \max\{k_b, S(b)\}$ and $S'(b') = S(b')$ for any $b' \in B$, $b' \neq b$.*

Lemma 3. *For any feasible seminar selections S and T, if for every seminar $b \in B$ the seminar selection $S \oplus (b, T(b))$ is feasible, then it holds that:*

$$\sum_{b \in B} [p(S \oplus (b, T(b))) - p(S)] \geq p(T) - p(S).$$

Proof. For a fixed SAP instance let G be its bipartite representation and let $G[V_S \cup I]$ and $G[V_T \cup I]$ be the bipartite representations of S and T respectively. Moreover, let M_S and M_T be two maximum weight matchings in $G[V_S \cup I]$ and $G[V_T \cup I]$ respectively. Remember that according to Lemma 1 it holds that $p(S) = \omega(M_S)$ and $p(T) = \omega(M_T)$. To prove the lemma we create matchings $\mathcal{M} = \{M_b\}_{b \in B}$ for the bipartite representations of assignments $p(S \oplus (b, T(b)))$, such that each edge of M_T is used in exactly one of the matchings in $\mathcal{M}$ and each edge of M_S is used in exactly $|B| - 1$ of the matchings in $\mathcal{M}$.

Let $\mathcal{C}$ be the collection of isolated components formed by the union of the edges of M_S and M_T. Since both M_S and M_T are matchings in G, each element of $\mathcal{C}$ is a path or cycle in G. For every $b \in B$ let $\mathcal{P}_b = \{P \in \mathcal{C} : V(P) \cap V_b \cap (V(M_T) \setminus V(M_S)) \neq \emptyset\}$, where $V(P)$ denotes the vertices of component P (Fig. 2).

Claim 2. For any $a \neq b \in B$, $\mathcal{P}_a \cap \mathcal{P}_b = \emptyset$.

Proof of claim: To prove the claim, assume that there exist $P \in \mathcal{P}_a \cap \mathcal{P}_b$ for some $a \neq b \in B$. Then by definition there exist $v_a \in V_a$ and $v_b \in V_b$ such that $v_a, v_b \in V(P)$ and $v_a, v_b \notin V(M_S)$ and therefore v_a and v_b are the endpoints of the alternating path P. Since neither of the endpoints of the path belong to M_S, P must have an odd number of edges. However, because both endpoints of P belong to the same partition of the bipartite graph G, the path P must have an even number of edges, hence the claim holds by contradiction.   □

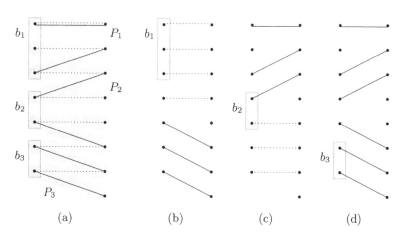

Fig. 2. An example with 3 seminars, b_1, b_2, b_3. (a) Two assignments M_S (dashed edges) and M_T (dotted edges); the three alternating paths formed by $M_S \cup M_T$ (light gray). $q(P_1) = b_1$ because it only intersects vertices from V_{b_1}; $q(P_2) = b_1$ because P_2 contains a vertex $V(M_T) \setminus V(M_S)$ that is in V_{b_1}; $r(P_3) = b_2$. (b), (c) and (d) assignments for seminar selections $S \oplus (b_1, 3)$, $S \oplus (b_2, 2)$ and $S \oplus (b_3, 2)$ combining edges of M_S and M_T.

Let $q : \mathcal{C} \to B$ be a map of the isolated components to the seminars with the following properties:

1. $q(P) \in \{b \in B : V(P) \cap V_b \neq \emptyset\}$;
2. if $P \in \mathcal{P}_b$ for any $b \in B$, $q(P) = b$.

Since $\mathcal{P}_b$ are disjoint by the previous claim and since for any seminar b it holds by definition that $V(P) \cap V_b \neq \emptyset$ whenever $P \in \mathcal{P}_b$, it is clear that such a mapping q exists.

For every $b \in B$ let M_b be the matching of G that uses all the edges of M_T from the alternating paths $P \in \mathcal{C}$ mapped by q to the seminar b, and all the edges of M_S from the paths $P \in \mathcal{C}$ mapped by q to some other seminar:

$$M_b = [M_T \cap E(q^{-1}(b))] \cup [M_S \cap (E(\mathcal{C}) \setminus E(q^{-1}(b)))].$$

Observe that any edge of M_T belongs to at least one matching M_b for some $b \in B$ and that any edge of M_S belongs to all but one of the matchings M_b. Therefore,

$$\sum_{b \in B} \omega(M_b) \geq \omega(M_T) + (|B| - 1) \cdot \omega(M_S).$$

Moreover, observe that for each $b \in B$, M_b is a matching in the bipartite representation of the seminar selection $S \oplus (b, T(b))$. Therefore $p(S \oplus (b, T(b))) = \omega(M_b)$ and the lemma follows. $\qquad \square$

Lemma 4. *Let S and T be two seminar selections such that $S \oplus (b, T(b))$ is feasible for every $b \in B$. Let $S^* = \arg\max_{S' \in inc(S)} (p(S') - p(S))/(c(S') - c(S))$. Then it holds that:*

$$\frac{p(S^*) - p(S)}{c(S^*) - c(S)} \geq \frac{p(T) - p(S)}{c(T)}.$$

Proof. By Lemma 3 we have that

$$\sum_{b \in B} [p(S \oplus (b, T(b))) - p(S)] \geq p(T) - p(S). \tag{2}$$

Since $\sum_{b \in B}[c(S \oplus (b, T(b))) - c(S)] \leq \sum_{b \in B} T(b) = c(T)$, inequality (2) implies that

$$\frac{\sum_{b \in B}[p(S \oplus (b, T(b))) - p(S)]}{\sum_{b \in B}[c(S \oplus (b, T(b))) - c(S)]} \geq \frac{p(T) - p(S)}{c(T)}. \tag{3}$$

Then, there exists at least one seminar $b^* \in B$ such that

$$\frac{p(S \oplus (b^*, T(b^*))) - p(S)}{c(S \oplus (b^*, T(b^*))) - c(S)} \geq \frac{p(T) - p(S)}{c(T)}. \tag{4}$$

Since $S \oplus (b^*, T(b^*))$ is clearly in $inc(S)$ the lemma follows directly from Eq. (4) and the definition of S^*. $\qquad \square$

Lemma 5. *Let T be a feasible seminar selection and let $r \in \mathbb{N}$ be such that $S_i \oplus (b, T(b))$ is feasible for every $i < r$ and $b \in B$. Then for each $i \leq r$ the following holds:*

$$p(S_i) - p(S_0) \geq \left[1 - \prod_{k=0}^{i-1} \left(1 - \frac{c(S_{k+1}) - c(S_k)}{c(T)} \right) \right] \cdot \left(p(T) - p(S_0) \right).$$

Proof. We prove the lemma by induction on the iterations i. By the definition of the algorithm, S_1 is the seminar selection with maximum marginal density in $inc(S_0)$, and thus Lemma 4 shows that the inequality holds for $i = 1$. Suppose that the lemma holds for iterations $1, ..., i$. We show that it also holds for iteration $i + 1$. For ease of exposition, for the remainder of this proof let $\alpha_i = \frac{c(S_{i+1}) - c(S_i)}{c(T)}$.

$$p(S_{i+1}) - p(S_0) = p(S_i) - p(S_0) + p(S_{i+1}) - p(S_i)$$
$$\geq p(S_i) - p(S_0) + \alpha_i \cdot (p(T) - p(S_i))$$
$$= (1 - \alpha_i)p(S_i) + \alpha_i \cdot p(T) - p(S_0)$$
$$\geq (1 - \alpha_i) \cdot \left(1 - \prod_{k=0}^{i-1}(1 - \alpha_k)\right)(p(T) - p(S_0))$$
$$+ (1 - \alpha_i) \cdot p(S_0) + \alpha_i \cdot p(T) - p(S_0)$$
$$= \left(1 - \alpha_i - \prod_{k=0}^{i}(1 - \alpha_k)\right)(p(T) - p(S_0))$$
$$+ \alpha_i \cdot (p(T) - p(S_0))$$
$$= \left(1 - \prod_{k=0}^{i}(1 - \alpha_k)\right)(p(T) - p(S_0)).$$

Where the first inequality follows from Lemma 4 and the second inequality follows from the induction hypothesis. □

Theorem 4. *When S_0 is the empty assignment the* **Greedy** *algorithm is a $\frac{1}{2} \cdot (1 - e^{-1})$ approximation for SAP.*

Proof. Let OPT be the seminar selection of a fixed optimal assignment solution for the given SAP instance. Let $b^* \in B$ be the seminar that is allocated the most students in OPT and let OPT' be the seminar selection for which $OPT'(b^*) = 0$ and $OPT'(b) = OPT(b)$ for any $b \neq b^* \in B$. Let r be the first iteration of the algorithm for which $c(S_r) > c(OPT')$. Clearly, $S_i \oplus (b, OPT(b))$ is feasible for every $i < r$ and $b \in B$. Since $p(S_0) = 0$, by applying Lemma 5 to iteration r we obtain:

$$p(S_r) \geq \left[1 - \prod_{k=0}^{r-1}\left(1 - \frac{c(S_{k+1}) - c(S_k)}{c(OPT')}\right)\right] \cdot p(OPT')$$
$$\geq \left[1 - \prod_{k=0}^{r-1}\left(1 - \frac{c(S_{k+1}) - c(S_k)}{c(S_r)}\right)\right] \cdot p(OPT'). \tag{5}$$

Observe that $c(S_r) = \sum_{k=0}^{r-1} c(S_{k+1}) - c(S_k)$ and that for any real numbers $a_0, ..., a_{r-1}$ with $\sum_{k=0}^{r-1} a_k = A$ it holds that:

$$\prod_{k=0}^{r-1}\left(1 - \frac{a_k}{A}\right) \leq \left(1 - \frac{1}{r}\right)^r < e^{-1}. \tag{6}$$

Therefore Eq. (5) implies $p(S_r) > (1 - e^{-1}) \cdot p(OPT')$. Since the profit of $\mathcal{A}_2$ is at least $p(b^*, OPT(b^*))$ it holds that

$$\mathcal{A}_1 + \mathcal{A}_2 > (1 - e^{-1}) \cdot p(OPT') + p(b^*, OPT(b^*))$$
$$\geq (1 - e^{-1}) \cdot p(OPT)$$

and therefore either $\mathcal{A}_1$ or $\mathcal{A}_2$ has profit at least $\frac{1}{2} \cdot (1 - e^{-1})p(OPT)$. □

5   Improving the Approximation

In this section we show that the algorithm can be improved by starting the greedy algorithm not from an empty seminar selection, but from a seminar selection that is part of the optimal solution. The improved algorithm is less efficient but achieves the optimal approximation ratio of $(1 - e^{-1})$. Let $\mathcal{A}_{opt}$ be an optimal seminar assignment and for any $b \in B$ let $p_{opt}(b)$ be the profit obtained in this assignment from seminar b:

$$p_{opt}(b) = \sum_{i \in \mathcal{A}_{opt}^{-1}(b)} p(i, b).$$

Clearly, the profit of the optimal solution is $\sum_{b \in B} p_{opt}(b)$. W.l.o.g, let b_1, b_2, b_3 be the three seminars of the optimal solution with highest profit and let S^* be a seminar selection such that $S^*(b) = OPT(b)$ if $b \in \{b_1, b_2, b_3\}$, and $S^*(b) = 0$ otherwise.

Theorem 5. *When* $S_0 = S^*$ *the* **Greedy** *algorithm is a* $\left(1 - e^{-1}\right)$*-approximation for SAP.*

Proof. Let OPT be the seminar selection corresponding to $\mathcal{A}_{opt}$. Let b^* be the seminar that is allocated the most students in OPT and is not allocated students in S^*. Moreover, let OPT' be the seminar selection for which $OPT'(b^*) = 0$ and $OPT'(b) = OPT(b)$ for any $b \neq b^* \in B$. Let r be the first iteration of the algorithm for which $c(S_r) > c(OPT')$. Clearly, the seminar selection $S_i \oplus (b, OPT(b))$ is feasible for every $i < r$ and $b \in B$. By applying Lemma 5 to iteration r we obtain:

$$p(S_r) - p(S^*) \geq \left[1 - \prod_{k=0}^{r-1}\left(1 - \frac{c(S_{k+1}) - c(S_k)}{c(OPT')}\right)\right] \cdot \left(p(OPT') - p(S^*)\right)$$

$$\geq \left[1 - \prod_{k=0}^{r-1}\left(1 - \frac{c(S_{k+1}) - c(S_k)}{c(S_r)}\right)\right] \cdot \left(p(OPT') - p(S^*)\right).$$

By applying Eq. (6) we obtain that

$$p(S_r) - p(S^*) \geq (1 - 1/e) \cdot \left(p(OPT') - p(S^*)\right),$$

and therefore

$$p(S_r) \geq (1 - 1/e) \cdot p(OPT') + p(S^*)/e$$
$$\geq (1 - 1/e) \cdot p(OPT) - p_{opt}(b^*) + p(S^*)/e. \tag{7}$$

By hypothesis S^* selects the three seminars with maximum profit in the optimal assignment and allocates exactly as many students to each as OPT does. Then, since $p_{opt}(b^*) \leq p_{opt}(b_i)$ for $i = 1, ..., 3$ it holds that $p(S^*) \geq 3 \cdot p_{opt}(b^*) > e \cdot p_{opt}(b^*)$ and the theorem follows. $\qquad \square$

Observe that the number of feasible seminar selections assigning students to at most three seminars is polynomial in the size of the input. Therefore, by repeatedly calling the greedy algorithm with all possible such selections our main result follows:

Corollary 1. *There exists a polynomial time* $(1 - e^{-1})$*-approximation algorithm for SAP.*

References

1. Bender, M., Thielen, C., Westphal, S.: A constant factor approximation for the generalized assignment problem with minimum quantities and unit size items. In: Mathematical Foundations of Computer Science, pp. 135–145 (2013)
2. Bender, M., Thielen, C., Westphal, S.: Erratum: A constant factor approximation for the generalized assignment problem with minimum quantities and unit size items. In: Mathematical Foundations of Computer Science, pp. E1–E3 (2013)
3. Cattrysse, D., Van Wassenhove, L.: A survey of algorithms for the generalized assignment problem. Eur. J. Oper. Res. **60**(3), 260–272 (1992)
4. Cohen, R., Katzir, L., Raz, D.: An efficient approximation for the generalized assignment problem. Inf. Process. Lett. **100**(4), 162–166 (2006)
5. Conforti, M., Cornuéjols, G.: Submodular set functions, matroids and the greedy algorithm: tight worst-case bounds and some generalizations of the Rado-Edmonds theorem. Discrete Appl. Math. **7**(3), 251–274 (1984)
6. Fleischer, L., Goemans, M., Mirrokni, V., Sviridenko, M.: Tight approximation algorithms for maximum general assignment problems. In: Proceedings of the Seventeenth Annual ACM-SIAM Symposium on Discrete Algorithm, SODA 2006 (2006)
7. Khuller, S., Moss, A., Naor, J.: The budgeted maximum coverage problem. Inf. Process. Lett. **70**(1), 39–45 (1999)
8. Krumke, S., Thielen, C.: The generalized assignment problem with minimum quantities. Eur. J. Oper. Res. **228**(1), 46–55 (2013)
9. Kundakcioglu, O., Alizamir, S.: Generalized assignment problem. In: Floudas, C., Pardalos, P. (eds.) Encyclopedia of Optimization, pp. 1153–1162. Springer, US (2009)
10. Nemhauser, G., Wolsey, L.: Maximizing submodular set functions: formulations and analysis of algorithms. In: Hansen, P. (ed.) Studies on Graphs and Discrete Programming, pp. 279–301. North-Holland, Amsterdam (1981)
11. Shmoys, D., Tardos, É.: An approximation algorithm for the generalized assignment problem. Math. Program. **62**(3), 461–474 (1993)
12. Sviridenko, M.: A note on maximizing a submodular set function subject to a knapsack constraint. Oper. Res. Lett. **32**(1), 41–43 (2004)
13. Vohra, R., Hall, N.: A probabilistic analysis of the maximal covering location problem. Discrete Appl. Math. **43**(2), 175–183 (1993)

A *priori* TSP in the Scenario Model

Martijn van Ee[1]([✉]), Leo van Iersel[2], Teun Janssen[2], and René Sitters[1,3]

[1] Vrije Universiteit Amsterdam, Amsterdam, The Netherlands
`{m.van.ee,r.a.sitters}@vu.nl`
[2] Delft University of Technology, Delft, The Netherlands
`{l.j.j.vaniersel,t.m.l.janssen}@tudelft.nl`
[3] Centrum voor Wiskunde en Informatica (CWI), Amsterdam, The Netherlands
`r.a.sitters@cwi.nl`

Abstract. In this paper, we consider the *a priori* traveling salesman problem (TSP) in the scenario model. In this problem, we are given a list of subsets of the vertices, called *scenarios*, along with a probability for each scenario. Given a tour on all vertices, the resulting tour for a given scenario is obtained by restricting the solution to the vertices of the scenario. The goal is to find a tour on all vertices that minimizes the expected length of the resulting restricted tour. We show that this problem is already NP-hard and APX-hard when all scenarios have size four. On the positive side, we show that there exists a constant-factor approximation algorithm in three restricted cases: if the number of scenarios is fixed, if the number of missing vertices per scenario is bounded by a constant, and if the scenarios are nested. Finally, we discuss an elegant relation with an *a priori* minimum spanning tree problem.

Keywords: Traveling salesman problem · A priori optimization · Master tour · Optimization under scenarios

1 Introduction

In universal and *a priori* routing, we extend our classical routing problems to the case that the set of clients is uncertain or changes regularly. Because reoptimizing over and over again might be inconvenient or impossible, we want to find a single tour. Given a tour and a set of clients, the active set, we shortcut the tour to the active set. In universal routing, the goal is to minimize the worst-case ratio of the value of the obtained solution and the deterministic optimal value. In *a priori* routing, we want to be good on average. The problem we consider in this paper is formally defined as follows.

In the *a priori* traveling salesman problem in the scenario model, we are given a complete weighted graph $G = (V, E)$ and a set of scenarios $\mathcal{S}$ with $S_1, \ldots, S_m \subseteq V$. Scenario S_j has probability p_j of being the active set, where $\sum_j p_j = 1$. We begin by finding an ordering on V, called the first-stage tour. When an active set is released, the second-stage tour is obtained by shortcutting the first-stage tour on the vertices of the active set. The goal is to find a first-stage tour that minimizes the expected length of the second-stage tour.

© Springer International Publishing AG 2017
K. Jansen and M. Mastrolilli (Eds.): WAOA 2016, LNCS 10138, pp. 183–196, 2017.
DOI: 10.1007/978-3-319-51741-4_15

This problem has, for example, a direct application to the photo-lithography processes used in semi-conductor manufacturing to transfer the geometric pattern of a chip onto a wafer [1]. This is done by putting UV light through a photomask on a photoresistant layer on top of the wafer. The entire wafer is not exposed at once, but one square at a time. If certain parts of the square do not need to be exposed, blades are moved in to block the UV light. Moving the blades is a time-consuming, and hence costly, process. Since it often influences the total processing time of a wafer in the lithography machine, minimizing the distance reduces the processing time. The blading positions are defined in a file. The blading positions are obtained from this file by reading it from top to bottom and the positions are used by the machine in order of appearance. A product will visit the photolithography machine multiple times during its fabrication. Every time it will use the same file that defines its blading positions, but it will not use all blading positions defined in the file in every visit. For each visit, there is a given subset of the blading positions that has to be used. Hence minimizing the movement of the blades comes down to finding an ordering of the blading positions such that the sum over all visits of the total distance between the blading-positions is minimized.

A priori TSP has already been considered in the independent decision and black-box model. In the independent decision model, vertex i is active with probability p_i, independent of the other vertices. Shmoys and Talwar [2] showed that a sample-and-augment approach gives a randomized 4-approximation, which can be derandomized to an 8-approximation algorithm. This factor was improved by van Zuylen [3] to 6.5. In the black-box model, we have no knowledge on the probability distribution over the vertices, but we are able to sample from it, i.e. to query the probability of any subset of the vertices. Schalekamp and Shmoys [4] showed that one can obtain a randomized $O(\log n)$-approximation even without sampling. A deterministic $O(\log^2 n)$-approximation can be obtained by using the result for universal TSP [5]. It was shown by [6] that there is an $\Omega(\log n)$ lower bound for deterministic algorithms on general metrics. By using the result of [5] and Theorem 3 in [6], we also know that there is no deterministic algorithm with guarantee $o\left(\sqrt[6]{\log n / \log \log n}\right)$ for planar metrics. For randomized algorithms, no lower bound is known for the black-box model.

The scenario model may be relevant for applications where the vertices are not active independently, but we do have some knowledge on the distribution. The former results give us the first results for *a priori* TSP in the scenario model. First of all, we inherit the randomized $O(\log n)$-approximation. Secondly, we know that a deterministic algorithm that does not use the information given in the scenarios will not achieve an approximation guarantee better than $O(\log n)$. The main question is whether we can use the scenarios to improve upon the $O(\log n)$ upper bound and which restrictions we can put on the scenarios in order to obtain constant-factor approximability.

The scenario model has not been studied extensively for other optimization problems. Immorlica et al. [7] investigated stochastic versions of Vertex Cover and Shortest Path. Ravi and Sinha [8] also looked at these problems and also defined stochastic scenario versions of Bin Packing, Facility Location and Set Cover. The problems in [8] differ from our setting in the sense that the weights used in the instance differ between scenarios. On the other hand, the work of [9] investigates a two-stage stochastic scheduling problem, where the set of jobs to be processed is uncertain. Finally, in [10], the classical scheduling problem of minimizing the makespan on two machines is considered in the *a priori* model with scenarios. It would be interesting to consider other stochastic combinatorial optimization problems in this framework.

In this paper, we will first examine the most natural lower bound, called the master tour lower bound. We use this lower bound to show that there exists a constant factor approximation algorithm for the problem if the number of scenarios is fixed. However, we also show that this lower bound cannot be used to improve upon the $O(\log n)$-approximation. We then look at several natural restrictions on the scenarios, namely small, big and nested. For small scenarios, we give strong inapproximability results. After that, we analyze the performance of the optimal tour on V for big scenarios. For nested scenarios, we show that there exists a 9-approximation algorithm. Finally, we show that there exists an elegant connection to an *a priori* minimum spanning tree problem. We end with a discussion on some open problems.

2   Master Tour Lower Bound

In this section, we explore the master tour lower bound. Here, we use that the contribution of scenario S_j to the objective value of an optimal solution, denoted by OPT, is at least $p_j T_j^*$, where T_j^* is the length of the optimal tour on S_j, so OPT $\geq \sum_j p_j T_j^*$. Two natural algorithms for *a priori* TSP in the scenario model are as follows. For each scenario, find an α-approximate tour, where α is the best approximation ratio available for TSP, and sort the scenarios on their resulting tour lengths T_j. Rename the scenarios such that $T_1 \leq T_2 \leq \ldots \leq T_m$. Now traverse the tours $1, 2, \ldots, m$, skipping already visited vertices, resulting in tour τ_1. Alternatively, rename the scenarios such that $p_1 \geq p_2 \geq \ldots \geq p_m$ and traverse the tours $1, 2, \ldots, m$, skipping already visited vertices, resulting in tour τ_2. We get the following result.

Theorem 1. *Tours τ_1 and τ_2 are $(2m - 1)$-approximations for a priori TSP in the scenario model, where $m \geq 2$ is the number of scenarios.*

Proof. Let us analyze tour τ_1. Consider an arbitrary scenario S_j. If D_j is the diameter of G restricted to S_j, we have $T_j^* \geq 2D_j$. Note that when analyzing the contribution of scenario S_j, it might happen that two tours, say T_x and T_y, with $x, y < j$, $S_x \cap S_j \neq \emptyset$ and $S_y \cap S_j \neq \emptyset$, belong to disjoint scenarios. In this case, we have to go from T_x to T_y. If $d(A, B)$ denotes the maximum distance between a vertex in A and a vertex in B, then this move costs us at most an extra $d(S_x \cap S_j, S_y \cap S_j)$. In the worst case, all scenarios before S_j have a non-empty intersection with S_j. For $j = 1$, the contribution is just $p_1 T_1 \leq \alpha p_1 T_1^*$. For $j \geq 2$, the contribution of S_j to the objective value of our solution is at most

$$p_j(T_1 + d(S_1 \cap S_j, S_2 \cap S_j) + T_2 + \ldots + d(S_{j-2} \cap S_j, S_{j-1} \cap S_j) + T_{j-1} + T_j)$$
$$\le p_j(jT_j + (j-2)D_j) \le p_j(\alpha j T_j^* + (j-2)\frac{1}{2}T_j^*) = ((\alpha + \frac{1}{2})j - 1)p_j T_j^*.$$

Note that you do not have to incur an extra distance from S_{j-1} to S_j, since they have a non-empty intersection. In general, this holds for the last scenario that intersects with S_j. The objective value is at most

$$\alpha p_1 T_1^* + \sum_{j=2}^{m}((\alpha + \frac{1}{2})j - 1)p_j T_j^* \le \left(\left(\alpha + \frac{1}{2} \right) m - 1 \right) \text{OPT}.$$

Since $\alpha = 1.5$ [11], we get a $2m - 1$-approximation algorithm. The analysis for τ_2 is similar.   □

It turns out that the master tour lower bound will not give a constant approximation for general metrics. This can be deduced from Theorem 2 in [6], which roughly states the following. Suppose you are given a d-regular Ramanujan graph G on n vertices with girth $g \ge \frac{2}{3}\log_{d-1} n$. Take a random walk of length $70g$ in G and let S be the vertices visited in this walk. Now, fix a first-stage tour. Theorem 2 in [6] states that for each of the first $g/2$ steps of the tour restricted to S, the probability that the edge has length $\Omega(\log n)$ is bounded from below by a constant.

Theorem 2. *There is an instance such that* $\text{OPT} = \Omega(\log n)\sum_j p_j T_j^*$ *and* $\text{OPT} = \Omega(\log m)\sum_j p_j T_j^*$.

Proof. As before, suppose you are given a d-regular Ramanujan graph G on n vertices with girth $g \ge \frac{2}{3}\log_{d-1} n$. The scenarios correspond to vertex sets induced by random walks of length $70g$ in G. For a fixed first-stage tour, Theorem 2 in [6] states that in each of the first $g/2$ steps of the second-stage tour, there is a constant fraction of the scenarios that use an edge of length $\Omega(\log n)$. This implies that the expected length of the first $g/2$ steps of the tour have expected length $\Omega(\log n)$. Since this is true for a constant number of steps, the lower bound also holds for the entire tour. Hence, we have an instance such that $\text{OPT} = \Omega(\log n)\sum_j p_j T_j^*$. The number of scenarios is equal to the number of possible walks of length $70g$. This is equal to $n \cdot d^{70g} = O(nd^{\log n}) = O(n^{\log d})$. Since d is a constant, this number is polynomially bounded. Hence, we have $\Theta(\log m) = \Theta(\log n)$, which gives us the second lower bound.   □

A similar question one can ask is whether a given instance has an optimal value that is equal to the master tour lower bound. Stated differently, is there a tour such that if we shortcut on the vertices of a scenario, we get the optimal solution for that scenario? This problem is known as the Master Tour problem. In the original problem, every subset of vertices was a scenario. Deineko et al. [12] showed that this problem is polynomially solvable. We can reformulate the problem to the case where we are given a set of scenarios and we only have to be optimal for these scenarios. It turns out that this problem is Δ_2^p-complete [13].

3   Small Scenarios

We start with showing that *a priori* TSP is still NP-complete when all scenarios are very small. We reduce from the Max Cut problem [14]. Here, we are given a graph $G = (V, E)$ and our goal is to find a set $S \subseteq V$ such that $|\delta(S, \bar{S})|$ is maximized, where $\delta(A, B)$ is the set of edges in the cut separating A from B.

Theorem 3. *A priori TSP is NP-complete when $|S_j| \leq 4$ for all j.*

Proof. We are given an instance of Max Cut. Create an instance of *a priori* TSP by making a complete graph G' on $V \cup \{s, t\}$. All edges with s or t as endpoint, except edge (s, t), have length 1 and all other edges have length 2 (see Fig. 1). For every edge (a, b) in E, we create a scenario $\{a, b, s, t\}$. All scenarios have equal probability. Note that a scenario can only have a contribution to the objective value of 4 or 6. We say that a scenario is satisfied if its resulting tour has length 4. Hence, minimizing the expected length is equivalent to maximizing the number of satisfied scenarios.

Suppose there is a cut of size at least k in G, say (Q_1, Q_2). First, visit the vertices of Q_1 in arbitrary order. After that, we visit s. Finally, we visit the vertices of Q_2 in arbitrary order followed by t. It is easy to see that every scenario corresponding to an edge in the cut has length 4, whereas other scenarios have length 6. Hence, there is a tour satisfying at least k scenarios.

On the other hand, suppose that we have a tour in G' satisfying at least k scenarios. Without loss of generality, the tour can be written as sR_1tR_2, where R_1 and R_2 are sequences of vertices. The only way to satisfy a scenario is by putting one vertex in R_1 and one vertex in R_2. Hence, the k satisfied scenarios correspond to edges in the cut (R_1, R_2) which has size at least k.   □

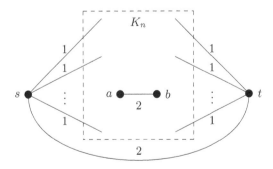

Fig. 1. Graph G' as in the proof of Theorem 3.

By adjusting the proof of Theorem 3, we can prove that the master tour problem with scenarios is NP-complete when $|S_j| \leq 5$. This is done by reducing from Set Splitting instead of Max Cut. The fact above follows because 3-Set

Splitting is NP-complete [15]. The master tour problem with scenarios is still open for $|S_j| \leq 4$.

Note that the graph we used in the proof above can be obtained by taking the metric completion of $K_{2,n}$. This graph is planar, bipartite and it has treewidth and pathwidth equal to 2. Deterministic TSP would be polynomially solvable on such a graph with bounded treewidth. Furthermore, there is a PTAS for deterministic TSP in planar graphs [16]. The next theorem shows that this is not the case for *a priori* TSP (since the proof uses the same graph as before, a metric completion of $K_{2,n}$). This theorem relies on the fact that Max Cut cannot be approximated within the Goemans-Williamson [17] constant, i.e. approximately 0.878567, unless the Unique Games Conjecture (UGC) fails [18], and it cannot be approximated within a factor $\frac{16}{17}$, unless P = NP [19].

Theorem 4. *There is no 1.0117-approximation for a priori TSP with $|S_j| \leq 4$, unless P = NP, and no 1.0242-approximation under UGC.*

Proof. Consider the reduction from the proof of Theorem 3. Let OPT_{TSP} and OPT_{CUT} be the optimal values of *a priori* TSP in the created instance and of Max Cut in the original instance respectively. We have $\text{OPT}_{\text{TSP}} = 6|E| - 2\text{OPT}_{\text{CUT}}$. If we have an $(1 + \alpha)$-approximation algorithm, we get a tour with total length at most $(1 + \alpha)(6|E| - 2\text{OPT}_{\text{CUT}})$. This implies that there are at least $(1 + \alpha)\text{OPT}_{\text{CUT}} - 3\alpha|E|$ satisfied scenarios. These correspond to edges in the cut, hence we have

$$\text{Size of cut} \geq (1 + \alpha)\text{OPT}_{\text{CUT}} - 3\alpha|E|$$
$$\geq (1 + \alpha)\text{OPT}_{\text{CUT}} - 6\alpha\text{OPT}_{\text{CUT}}$$
$$= (1 - 5\alpha)\text{OPT}_{\text{CUT}},$$

where the second inequality follows from $\text{OPT}_{\text{CUT}} \geq |E|/2$. Assuming P $\neq$ NP or the Unique Games Conjecture, this means that there is no $(1+\alpha)$-approximation for our problem for α's with $1 - 5\alpha \geq \frac{16}{17}$ or $1 - 5\alpha \geq 0.878567$ respectively. These inequalities are tight for $\alpha \approx 0.0117$ and $\alpha \approx 0.0242$ respectively.   □

One could also consider the path-version of *a priori* TSP. In fact, the application on photolithography is modeled as the path-version. It is easy to see that this problem is trivial when $|S_j| \leq 2$ for all j. If we delete t from the graph created in the reduction of Theorem 3, we can use this graph and the same reduction to show that the path-version of *a priori* TSP is NP-complete when $|S_j| \leq 3$. It is easy to see that this graph can be obtained by taking the metric completion of the star graph. Since the star has pathwidth 2, the problem is NP-complete on graphs with this property. On the other hand, the problem is trivially solvable on path graphs. Note that we can also adjust Theorem 4 to the path-version which will give the same inapproximability result, i.e. there is no 1.0117-approximation, unless P = NP, and there is no 1.0242-approximation under UGC.

We can strengthen the inapproximability of *a priori* TSP by using strong results on Permutation CSP's [20]. The problem that we need will be called 4-Undirected Cyclic Ordering (4-UCO). In this problem, we are given a ground set T and a set of 4-tuples Δ using elements from T. Our goal is to construct an ordering on T that maximizes the number of satisfied 4-tuples. We say that 4-tuple (a, b, c, d) is satisfied if one of the following sequences is a subsequence of the total ordering: (a, b, c, d), (b, c, d, a), (c, d, a, b), (d, a, b, c), (d, c, b, a), (c, b, a, d), (b, a, d, c), (a, d, c, b). In other words, we get a collection of cycles and we want to find an ordering maximizing the number of cycles that can be embedded in it. To the best of our knowledge, the problem has never been considered. For completeness, we first show that the problem is NP-complete by using a reduction from Cyclic Ordering. In this problem, we are given a set of ordered triples Δ of ground set T. The question is whether there exists a cyclic ordering on all elements such that each triple is ordered in the right direction. This problem is NP-complete [21].

Theorem 5. *4-Undirected Cyclic Ordering is NP-complete.*

In [20], it is shown that every Permutation CSP of constant arity is approximation resistant. This means that, under the Unique Games Conjecture, the best we can do is constructing a random ordering. Classical problems like Cyclic Ordering and Betweenness are in this class of problems. It is easy to see that 4-UCO is also in this class. A corollary of the work of Guruswami et al. is that there is no approximation algorithm with guarantee greater than $\frac{1}{3}$, assuming the Unique Games Conjecture is true. The natural generalization of 4-UCO is 5-UCO. For this problem, there is no algorithm having a guarantee greater than $\frac{1}{12}$. This gives the following results.

Theorem 6. *Under UGC, there is no α-approximation for a priori TSP with*

(a) $\alpha < \frac{10}{9}$ *when* $|S_j| \leq 6$,
(b) $\alpha < \frac{7}{6}$ *when* $|S_j| \leq 8$,
(c) $\alpha < \frac{71}{60}$ *when* $|S_j| \leq 10$.

For the path-version, we can strengthen previous results by using the maximization version of Betweenness. In this problem, we are given a set of triples Δ from elements of T. The triple (a, b, c) is satisfied if (a, b, c) or (c, b, a) is a subsequence of the total ordering. The goal is to find an ordering on T maximizing the number of satisfied triples. By [20], the best approximation ratio is $\frac{1}{3}$, unless the Unique Games Conjecture fails. Under the assumption that P $\neq$ NP, there is no approximation for Max Betweenness with a factor better than $\frac{1}{2}$ [22].

Theorem 7. *There is no $\frac{9}{8}$-approximation for a priori path-TSP with $|S_j| \leq 5$, unless $P = NP$, and no $\frac{7}{6}$-approximation under UGC.*

Finally, we note that by using twice the diameter of a scenario as a lower bound, we can show that an arbitrary tour is a $c/2$-approximation when $|S_j| \leq c$. A random tour gives a value of at most $(c^2 - 3c + 4/2c - 2)$ times the optimal value in expectation. This factor approaches $c/2$ for c large. Similar results hold for the path-version.

4   Big Scenarios

In this section, we investigate the special case of big scenarios, i.e. the case when each scenario has size at least $n - c$, for small c. One would expect that the optimal tour on the entire instance would perform well on these instances. Here, we analyze this option. Let us denote $\text{OPT}(S)$ for the optimal value of a tour on S. Further, let $\text{OPT}(S)|_T$ denote the value of the optimal tour on S shortcutted to T. As before, let D_S denote the diameter of the graph restricted to S.

Lemma 1. *For $S \subset V$ and $1 \leq c \leq \lceil n/2 \rceil$ such that $|S| = n - c$, we have*

$$\text{OPT}(V)|_S \leq \text{OPT}(S) + cD_S.$$

Proof. Suppose $S = V \setminus \{a_1, \ldots, a_c\}$. Let $\Delta_S^{a_i} = \min_{u \in S} d(u, a_i)$ for $i = 1, \ldots, c$. Since we can extend our tour on S to V by going back and forth to each a_i, we have $\text{OPT}(V) \leq \text{OPT}(S) + 2\sum_{i=1}^c \Delta_S^{a_i}$. We want to show that $\text{OPT}(V)|_S \leq \text{OPT}(S) + cD_S$. Suppose this is not the case, i.e. $\text{OPT}(V)|_S > \text{OPT}(S) + cD_S$. Furthermore, suppose w.l.o.g. that b_i and d_i are the two nodes in S that are visited before and after a_i in the optimal tour of V. If two consecutive vertices on the tour are not in S, then one can reconstruct the tour accordingly without increasing the length of the tour restricted to S. Then

$$\text{OPT}(V) = \text{OPT}(V)|_S + \sum_{i=1}^c (d(b_i, a_i) + d(a_i, d_i) - d(b_i, d_i))$$

$$\geq \text{OPT}(V)|_S + \sum_{i=1}^c (2\Delta_S^{a_i} - d(b_i, d_i))$$

$$\geq \text{OPT}(V)|_S - cD_S + 2\sum_{i=1}^c \Delta_S^{a_i}$$

$$> \text{OPT}(S) + 2\sum_{i=1}^c \Delta_S^{a_i}$$

But this is a contradiction to our previous observation. Hence $\text{OPT}(V)|_S \leq \text{OPT}(S) + cD_S$. □

Theorem 8. *The optimal solution on V is a $(1 + \frac{c}{2})$-approximation for a priori TSP with $|S_i| \geq n - c$, where $1 \leq c \leq \lceil \frac{n}{2} \rceil$.*

5   Nested Scenarios

Let us now consider the case of nested scenarios, i.e. $S_1 \subseteq S_2 \subseteq \ldots \subseteq S_m$. Here, the following algorithm gives a constant factor approximation. First, compute an 1.5-approximate tour T_j for scenario S_j for all j. Let $\alpha_1 = 1$. Next, for $h = 2, 3, \ldots$ let α_h be the largest number $k > \alpha_{h-1}$ for which $T_k \leq 2T_{\alpha_{h-1}}$. If no such k exists then let $\alpha_h = \alpha_{h-1} + 1$. The first-stage tour is obtained by visiting vertices in the order $T_{\alpha_1}, T_{\alpha_2}, \ldots$.

Theorem 9. *The algorithm above is a 9-approximation for nested scenarios.*

Proof. Consider scenario S_j. The last vertices of this scenario will be visited on the tour T_{α_h}, where h is the smallest index such that $\alpha_h \geq j$. Note that for any $h \geq 2$, we have $T_{\alpha_h} > 2T_{\alpha_{h-2}}$. Hence, we can decompose the concatenated tour up to T_{α_h} into two parts which correspond to even and odd h respectively, such that both parts have geometrically increasing tour lengths. The length of the concatenated tour up to T_{α_h} is therefore at most

$$2T_{\alpha_{h-1}} + 2T_{\alpha_h}.$$

If $\alpha_h = j$ then the length of the tour is at most $2T_{\alpha_{h-1}} + 2T_{\alpha_h} \leq 4T_{\alpha_h} = 4T_j \leq 6T_j^*$.

If $\alpha_h > j$ then we must have $T_{\alpha_h} \leq 2T_{\alpha_{h-1}}$ so the length of the tour is at most $2T_{\alpha_{h-1}} + 2T_{\alpha_h} \leq 6T_{\alpha_{h-1}} \leq 6T_j \leq 9T_j^*$. $\qquad\square$

The problem is still open for laminar scenarios, i.e. when for each i, j, either $S_i \cap S_j = \emptyset$ or $S_i \subseteq S_j$ or $S_j \subseteq S_i$. It is even open in the case when the scenarios have the following structure.

$$S_i \cap S_j = \emptyset \text{ for } i \neq j, i, j = 1, \ldots, m-1, \text{ and } S_m = \bigcup_{j=1}^{m-1} S_j.$$

It would be interesting if one could get a constant factor approximation for these "starlike" (the inclusion graph is a star) instances.

6   Relation with Minimum Spanning Tree Problems

It would be nice to have a similar relation between *a priori* TSP and *a priori* MST as in the deterministic setting. We consider two versions of *a priori* MST. The first one is defined by Bertsimas [23], who called it *a priori* MST, while it seems more natural to call it *a priori* Steiner Tree. The second problem is defined by Boria et al. [24], who called it Probabilistic MST under Closest Ancestor. In both problems, we have a graph $G = (V, E)$ and a probability distribution over subsets of vertices. The second problem also has a root r that is always active. This is optional in the first problem. The goal is to construct a tree on the entire vertex set in the first stage. A subset S of the vertices, drawn according to the probability distribution, is revealed in the second stage. In the *a priori* MST, the second-stage tree will be obtained by deleting inactive vertices, provided that the remaining tree stays connected. In the Probabilistic MST under Closest Ancestor, the second-stage tree only contains active vertices. This is done by taking an edge between an active vertex and its closest active ancestor in the rooted first-stage tree. In both problems, the goal is to construct a first-stage tour that minimizes the expected length of the second-stage tree.

Unfortunately, it turns out that the expected length of the optimal *a priori* MST defined by Bertsimas is not smaller than the optimal *a priori* TSP in general. The gap between the optimal values of *a priori* MST and *a priori* TSP can be arbitrarily large.

Theorem 10. *The optimal value of the a priori MST can be arbitrarily greater than the optimal value of the a priori TSP.*

However, the Probabilistic MST under Closest Ancestor can be used as a lower bound for *a priori* TSP. In fact, we only lose a factor 2. Note that this only works for the rooted case, since Probabilistic MST under Closest Ancestor is defined with a root vertex.

Theorem 11. *If there is an α-approximation for the Probabilistic MST under Closest Ancestor, then there is a 2α-approximation for the a priori TSP, and vice versa.*

Proof. We show that the following inequalities are valid, where OPT_{MST} and OPT_{TSP} denote the optimal values of Closest Ancestor and *a priori* TSP respectively.

$$\text{OPT}_{\text{MST}} \le \text{OPT}_{\text{TSP}} \le 2\text{OPT}_{\text{MST}} \le 2\text{OPT}_{\text{TSP}}.$$

The first inequality can be proven by taking the optimal *a priori* TSP-tour and deleting one edge. This gives a spanning tree on V, called T. If we look at a specific active set S, then the optimal *a priori* TSP-tour restricted to S will have exactly on edge less than before. Namely, if we delete edge (a, b) from tour $(1, \ldots, a, b, \ldots, n)$, only edge $(\max\{k \in S : k \le a\}, \min\{k \in S : k \ge b\})$ will disappear from the restricted tour on S. Note that for active set S, the tour without this edge is the same as T shortcutted to S. Hence, this is a feasible solution for Probabilistic MST under Closest Ancestor with cost no larger than the optimal value of *a priori* TSP, and the first inequality has been proven.

The second inequality is proven by doubling the optimal tree and shortcutting the obtained Eulerian tour. In each scenario, the cost of the edges is at most twice the cost of the edges in the tree restricted to the scenario. The third inequality follows from the first inequality. ∎

Corollary 1. *There is a randomized 8-approximation and a deterministic 13-approximation for Probabilistic MST under Closest Ancestor in the independent decision model. There is also a $O(\log n)$-approximation in the black-box model.*

Unfortunately, this does not imply a 2-approximation for *a priori* TSP, since we can prove that Probabilistic MST under Closest Ancestor is NP-complete in the scenario model. For this, we need the following lemma.

Lemma 2. *If Probabilistic MST under Closest Ancestor is NP-complete in the non-metric case, then it is NP-complete in the metric case.*

Boria et al. [24] showed that Probabilistic MST under Closest Ancestor is NP-complete in the independent decision model, but only for the non-metric case. Using Lemma 2, we obtain the following corollary.

Corollary 2. *Probabilistic MST under Closest Ancestor is NP-complete in the independent decision model, even if the triangle inequality is satisfied.*

Theorem 12. *Probabilistic MST under Closest Ancestor in the scenario model is NP-complete.*

Proof. We reduce the problem from the NP-complete problem Exact Cover by 3-Sets [14]. In this problem, we are given $3q$ elements, $x_1, \ldots, x_{3q}$, and m sets, $y_1, \ldots, y_m$, containing three elements. The problem asks whether there are q sets that together cover all elements. Create the graph as in Fig. 2. Non-present edges have weight equal to M, where M is a large number. There are m scenarios with probability $1/m$. In each scenario, all x_j's, r and s are active as is one of the y_i's.

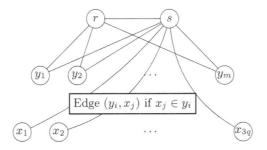

Fig. 2. Graph used in proof of Theorem 12. Edges (r, s) and (r, y_i) have length 0. Edges (s, y_i) and (y_i, x_j) have length 1. Edges (s, x_j) have length 2. Non-drawn edges have length M.

If there is an exact cover, then construct the following solution. If set y_i is chosen in the cover, then use edge (s, y_i) and the edges from vertex y_i to the corresponding elements of y_i. If set y_i is not in the cover, then use edge (r, y_i). Finally, use edge (r, s). This solution has expected value equal to $q(1/m \cdot 4 + (m-1)/m \cdot 6) = q(6 - 2/m)$.

Note that an optimal tree will never use edges with weight M or a combination of edges that enforce using an edge of weight M in the shortcut solution. This leaves five ways of connecting a specific set vertex y_i and element vertex x_j, where j is in set i, to r and s. The five subtrees are depicted in Fig. 3.

Tree T_3 is dominated by T_1, since T_1 only has cost 2 for connecting x_j when y_i is inactive while T_3 always has cost 2. Similarly, T_4 is dominated by T_2 and T_5 is dominated by T_1. So, an optimal tree is a combination of T_1 and T_2. Suppose that the tree connects k set vertices to s which connect ℓ elements vertices. The other set vertices are connected to r whereas the other element vertices are connected to s. Number the k set vertices connected to s as $1, \ldots, k$ and say that set vertex i connects ℓ_i element vertices. This tree has an expected value of

$$\frac{1}{m} \sum_{i=1}^{k} ((\ell_i + 1) - 2(3q - \ell_i)) + \frac{m-k}{m} 6q = 6q + \frac{1}{m}(k - \ell),$$

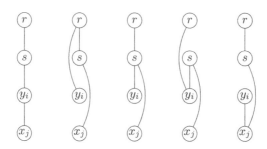

Fig. 3. Subtrees T_1 up to T_5.

which is equal to $q(6 - 2/m)$ if and only if $k = q$ and $\ell = 3q$. Hence, there is a tree with expected value at most $q(6 - 2/m)$ if and only if there is an exact cover. Using Lemma 2 completes the proof.   □

7   Conclusion

In this paper, we showed how to get constant factor approximation for some well-structured problem instances. An interesting question that remains unanswered is whether there exists a constant factor approximation for laminar scenarios. More specifically, it is still open whether we can do this on "starlike" scenarios. One could also consider instances with restricted metrics. In Sect. 3 we showed that there is no PTAS for planar graphs. We do not have such results in the Euclidean plane. It would be interesting to settle the approximability of the problem in this metric. It is easy to construct examples where the optimal solution crosses itself and hence the non-crossing property does not hold. This property was a crucial ingredient of the PTAS by Arora [25] for the deterministic problem. So far, we have not been able to show any lower bound or improve the upper bound for this special case.

We did not succeed in improving the $O(\log n)$-approximation for the general problem. Next to the master tour lower bound, we investigated minimum spanning tree and linear programming approaches. However, preliminary results suggest that these approaches might not help us to break the barrier. In fact, we conjecture that there is no $o(\log n)$-approximation algorithm for *a priori* TSP in the scenario model in the general case.

Acknowledgments. We would like to thank Karen Aardal, Jan Driessen and Neil Olver for useful discussions. A part of the work by Teun Janssen has been performed in the project INTEGRATE "Integrated Solutions for Agile Manufacturing in High-mix Semiconductor Fabs", co-funded by grants from France, Italy, Ireland, The Netherlands and the ECSEL Joint Undertaking. Martijn van Ee and René Sitters are supported by the NWO Grant 612.001.215. Leo van Iersel was partially supported by NWO and partially by the 4TU Applied Mathematics Institute.

References

1. Driessen, J., Janssen, T.: Minimizing the blading in lithography machines: an application of the *a priori* TSP problem. Unpublished manuscript (2016)
2. Shmoys, D., Talwar, K.: A constant approximation algorithm for the *a priori* traveling salesman problem. In: Lodi, A., Panconesi, A., Rinaldi, G. (eds.) IPCO 2008. LNCS, vol. 5035, pp. 331–343. Springer, Heidelberg (2008). doi:10.1007/978-3-540-68891-4_23
3. Zuylen, A.: Deterministic sampling algorithms for network design. Algorithmica **60**, 110–151 (2011)
4. Schalekamp, F., Shmoys, D.B.: Algorithms for the universal and *a priori* TSP. Oper. Res. Lett. **36**, 1–3 (2008)
5. Hajiaghayi, M.T., Kleinberg, R., Leighton, T.: Improved lower and upper bounds for universal TSP in planar metrics. In: Proceedings of the Seventeenth Annual ACM-SIAM Symposium on Discrete Algorithms, pp. 649–658 (2006)
6. Gorodezky, I., Kleinberg, R.D., Shmoys, D.B., Spencer, G.: Improved lower bounds for the universal and *a priori* TSP. In: Serna, M., Shaltiel, R., Jansen, K., Rolim, J. (eds.) APPROX/RANDOM 2010. LNCS, vol. 6302, pp. 178–191. Springer, Heidelberg (2010). doi:10.1007/978-3-642-15369-3_14
7. Immorlica, N., Karger, D.R., Minkoff, M., Mirrokni, V.S.: On the costs and benefits of procrastination: approximation algorithms for stochastic combinatorial optimization problems. In: Proceedings of the Fifteenth Annual ACM-SIAM Symposium on Discrete Algorithms, pp. 691–700 (2004)
8. Ravi, R., Sinha, A.: Hedging uncertainty: approximation algorithms for stochastic optimization. Math. Program. **108**, 97–114 (2006)
9. Chen, L., Megow, N., Rischke, R., Stougie, L.: Stochastic and robust scheduling in the cloud. In: Proceedings of the 18th International Workshop on Approximation Algorithms for Combinatorial Optimization Problems, pp. 175–186 (2015)
10. Feuerstein, E., Marchetti-Spaccamela, A., Schalekamp, F., Sitters, R., Ster, S., Stougie, L., Zuylen, A.: Scheduling over scenarios on two machines. In: Cai, Z., Zelikovsky, A., Bourgeois, A. (eds.) COCOON 2014. LNCS, vol. 8591, pp. 559–571. Springer, Heidelberg (2014). doi:10.1007/978-3-319-08783-2_48
11. Christofides, N.: Worst-case analysis of a new heuristic for the travelling salesman problem. Technical report, DTIC Document (1976)
12. Deineko, V.G., Rudolf, R., Woeginger, G.J.: Sometimes travelling is easy: the master tour problem. SIAM J. Discrete Math. **11**, 81–93 (1998)
13. Ee, M., Sitters, R.: On the complexity of master problems. In: Italiano, G.F., Pighizzini, G., Sannella, D.T. (eds.) MFCS 2015. LNCS, vol. 9235, pp. 567–576. Springer, Heidelberg (2015). doi:10.1007/978-3-662-48054-0_47
14. Karp, R.M.: Reducibility among combinatorial problems. In: Miller, R.E., Thatcher, J.W., Bohlinger, J.D. (eds.) Complexity of Computer Computations, pp. 85–103. Springer, Heidelberg (1972)
15. Lovász, L.: Coverings and colorings of hypergraphs. In: Proceedings of the 4th Southeastern Conference on Combinatorics, Graph Theory and Computing, pp. 3–12 (1973)
16. Arora, S., Grigni, M., Karger, D.R., Klein, P.N., Woloszyn, A.: A polynomial-time approximation scheme for weighted planar graph TSP. In: Proceedings of the 9th Annual ACM-SIAM Symposium on Discrete Algorithms, pp. 33–41 (1998)
17. Goemans, M.X., Williamson, D.P.: Improved approximation algorithms for maximum cut and satisfiability problems using semidefinite programming. J. ACM (JACM) **42**, 1115–1145 (1995)

18. Khot, S., Kindler, G., Mossel, E., O'Donnell, R.: Optimal inapproximability results for MAX-CUT and other 2-variable CSPs? SIAM J. Comput. **37**, 319–357 (2007)
19. Håstad, J.: Some optimal inapproximability results. J. ACM **48**, 798–859 (2001)
20. Guruswami, V., Håstad, J., Manokaran, R., Raghavendra, P., Charikar, M.: Beating the random ordering is hard: every ordering CSP is approximation resistant. SIAM J. Comput. **40**, 878–914 (2011)
21. Galil, Z., Megiddo, N.: Cyclic ordering is NP-complete. Theor. Comput. Sci. **5**, 179–182 (1977)
22. Austrin, P., Manokaran, R., Wenner, C.: On the NP-hardness of approximating ordering-constraint satisfaction problems. Theor. Comput. **11**, 257–283 (2015)
23. Bertsimas, D.: Probabilistic combinatorial optimization problems. Ph.D. thesis, Massachusetts Institute of Technology (1988)
24. Boria, N., Murat, C., Paschos, V.: On the probabilistic min spanning tree problem. J. Math. Model. Algorithms **11**, 45–76 (2012)
25. Arora, S.: Polynomial time approximation schemes for Euclidean traveling salesman and other geometric problems. J. ACM (JACM) **45**, 753–782 (1998)

Local Search Based Approximation Algorithms for Two-Stage Stochastic Location Problems

Felix J.L. Willamowski[1(✉)] and Andreas Bley[2]

[1] Operations Research, RWTH Aachen University, 52072 Aachen, Germany
willamowski@or.rwth-aachen.de
[2] Institute of Mathematics, University of Kassel, 34132 Kassel, Germany
abley@mathematik.uni-kassel.de

Abstract. We present a nested local search algorithm to approximate several variants of metric two-stage stochastic facility location problems. These problems are generalizations of the well-studied metric uncapacitated facility location problem, taking uncertainties in demand values and costs into account. The proposed nested local search procedure uses three facility operations: adding, dropping, and swapping. To the best of our knowledge, this is the first constant-factor local search approximation for two-stage stochastic facility location problems.

Besides traditional direct assignments from clients to facilities, we also investigate shared connections via capacitated trees and tours. We obtain the first constant-factor approximation algorithms for both connection types in the setting of two-stage stochastic optimization. Our algorithms admit order-preserving metrics and thus significantly generalize and improve the allowed mutability of the metric in comparison to previous algorithms, which only allow scenario-dependent inflation factors.

1 Introduction

In this paper we study stochastic generalizations of the metric uncapacitated facility location (UFL) problem. The UFL problem was introduced in the early 1960's and is one of the most studied problems in the discrete optimization literature. The first constant-factor approximation algorithm for the metric case, where the assignment costs satisfy the triangle inequality, was presented in the late 1990's by Shmoys et al. [12]. From that time onward, many other constant-factor approximations have been developed, decreasing the approximation factor rapidly to 1.488, the currently best known proposed by Li [9]. Ye and Zhang [16] observed that so far each algorithm for approximating the metric UFL problem uses at least one of the following three paradigms: LP rounding, primal-dual, or local search techniques. LP rounding and primal-dual techniques were also applied to the two-stage stochastic version of the problem, but, to the best of our knowledge, no pure local search approaches have been used. One purpose of this paper is to close this gap, especially because local search turned out to be a powerful tool for approximating capacitated location problems. Moreover, the proposed local search approach allows more mutability of the metrics than previous approaches and it is very easy to implement in practice.

© Springer International Publishing AG 2017
K. Jansen and M. Mastrolilli (Eds.): WAOA 2016, LNCS 10138, pp. 197–209, 2017.
DOI: 10.1007/978-3-319-51741-4_16

The metric two-stage stochastic uncapacitated facility location (tsUFL) problem was introduced in 2004 by Ravi and Sinha [10]. It models the task of locating *facilities* to serve demands of *clients* as a two-stage stochastic optimization problem with recourse, where a set of scenarios depict the possible outcomes of the future. The decision making process, essentially deciding which facilities to open, is divided into two stages. In a first stage, decisions are made with incomplete knowledge about the future, i.e., only the probability distribution of the scenarios with their parameters is known. In a second (recourse) stage, information is revealed about which scenario is realized and additional recourse decisions are made. The goal is to minimize the fixed first-stage and the expected second-stage cost. There are two main concepts to express the probability distribution of the scenarios in the literature. In the *scenario model* each scenario with its parameters and its associated probability is explicitly given as part of the input. An assumption commonly made in this model is that the number of scenarios is polynomial bounded by the other input parameters (e.g., number of facilities and clients). In the *black-box model*, the probability distribution is only given implicitly by an algorithm that draws independent samples of the distribution. Although the black-box model is more general than the scenario model, Charikar et al. [3] were able to show that, under reasonable assumptions on the distribution and losing only a factor $(1 + O(\varepsilon))$ in the objective, the black-box model reduces to the scenario model with only a polynomial number of samples. For this reason, we only consider the scenario model.

In the tsUFL problem we assume that the facilities opened in the first stage are present in each scenario, whereas facilities opened in the second stage exist only for their specific scenario. For each scenario, the clients have to be served by either an open facility of the first stage or by a facility opened in the second stage for this specific scenario. The service costs form a metric. Clearly, the approximability depends on how much the metric varies over the scenarios. We will extend the (rather restrictive) concept of scenario-dependent inflation factors used in previous works to a more general scenario-dependent *mutable metric*. The currently best known approximation algorithm for tsUFL with inflation factors is given by Ye and Zhang [16] with a factor of 1.86.

Formally, an instance of the tsUFL problem with mutable metric is given by a complete graph $G = (V, E)$ on the node set $V = \mathcal{C} \cup \mathcal{F}$ of clients $\mathcal{C}$ and facilities $\mathcal{F}$, *first-stage facility opening costs* $f_i \in \mathbb{Q}_{\geq 0}$, $i \in \mathcal{F}$, and a set of m possible scenarios. For the sake of simplicity, we index the scenarios by $k \in [m] := \{1, \ldots, m\}$ and say scenario k instead of scenario indexed by k. Scenario k occurs with probability p_k and is defined by *second-stage facility opening costs* $f_i^k \in \mathbb{Q}_{\geq 0}$, $i \in \mathcal{F}$, a metric *service cost function* $c^k : E \to \mathbb{Q}_{\geq 0}$, and client demands $d_j^k \in \mathbb{Q}_{\geq 0}$, $j \in \mathcal{C}$. The goal is to find a set of *first-stage facilities* $F \subseteq \mathcal{F}$, which is independent of the realization of the scenario, and, for each scenario $k \in [m]$, a set of *second-stage facilities* $F^k \subseteq \mathcal{F}$ and an assignment $\sigma^k : \mathcal{C} \to F \cup F^k$, which minimize first-stage and expected second-stage costs

$$\sum_{i \in F} f_i + \sum_{k=1}^{m} p_k \cdot \Big(\sum_{i \in F^k} f_i^k + \sum_{j \in C} d_j^k \cdot c^k(\sigma^k(j), j) \Big).$$

In order to appropriately model problem variants where multiple clients may share parts of a network that connect them to the facilities, we introduce the two-stage stochastic facility location problem with tree-connections (tsUFL-T). Formally, this problem is defined as follows. The graph, the first-stage facility opening costs, and the set of m possible scenarios with their parameters are given as in the tsUFL problem. Additionally, let $C_+^k := \{ j \in C \mid d_j^k > 0 \}$ denote the set of clients with positive demand in scenario k. The goal is to find a set of first-stage facilities $F \subseteq \mathcal{F}$ and, for each scenario k, a set of second-stage facilities $F^k \subseteq \mathcal{F}$ and a set T^k of trees in $G[F \cup F^k \cup C_+^k]$ such that each tree contains exactly one facility, i.e., $|V(T) \cap (F \cup F^k)| = 1$ for all $T \in T^k$, and all clients with positive demand are served, i.e., $C_+^k \subseteq \bigcup_{T \in T^k} V(T)$, which minimize

$$\sum_{i \in F} f_i + \sum_{k=1}^{m} p_k \cdot \Big(\sum_{i \in F^k} f_i^k + \sum_{T \in T^k} \sum_{e \in E(T)} c^k(e) \Big).$$

As an intermediate step towards approximation algorithms for problems with capacitated trees and tours later in the paper, we first combine the connection types of tsUFL and tsUFL-T and study the two-stage stochastic uncapacitated facility location problem with direct and tree-connections (tsUFL-DT), where each client is served twice, directly and via a shared tree. This problem also may be of independent interest for some applications.

Since in many applications the connection network cannot handle unlimited amounts of flow, we examine capacitated network connection types like the metric two-stage stochastic capacitated-cable facility location (tsCCFL) problem. In this problem, we additionally need to select edge capacities that permit to route the clients' demands simultaneously to the open facilities. Formally, an instance of the tsCCFL problem is given by a complete graph $G = (V, E)$ with $\mathcal{F} \cup \mathcal{C} \subseteq V$. The first-stage facility opening costs and the set of scenarios with their parameters are defined as in the tsUFL problem. Additionally, there is a *cable capacity* $u \in \mathbb{Z}_{>0}$ limiting the demand flow. The task is to choose a set of first-stage facilities $F \subseteq \mathcal{F}$ and, for each scenario k, a set of second-stage facilities $F^k \subseteq \mathcal{F}$, a set T^k of trees in G such that each tree is rooted at an open facility and each client with positive demand is served, and a number of cables $z_e^k \in \mathbb{Z}_{\geq 0}$ for each edge $e \in \bigcup_{T \in T^k} E(T)$ such that the flow given by routing all demands simultaneously via the tree edges to the open facilities does not exceed the edge capacities $z_e^k \cdot u$. As before, we wish to minimize the expected costs

$$\sum_{i \in F} f_i + \sum_{k=1}^{m} p_k \cdot \Big(\sum_{i \in F^k} f_i^k + \sum_{T \in T^k} \sum_{e \in E(T)} z_e^k \cdot c^k(e) \Big).$$

As the second problem with a capacitated connection we consider the two-stage stochastic capacitated location routing (tsCLR) problem. It combines the tsUFL problem with the well-studied capacitated vehicle routing problem. Formally, an instance of tsCLR is given by a complete graph, first-stage facility opening costs, and a set of scenarios with parameters as in the tsUFL problem. Additionally, there is a *vehicle capacity* $u \in \mathbb{Z}_{>0}$. The task is to find a set of first-stage facilities $F \subseteq \mathcal{F}$ and, for each scenario k, a set of second-stage facilities $F^k \subseteq \mathcal{F}$, a set of tours $\mathcal{T}^k$ with *demand assignment* $x^k : \mathcal{C} \times \mathcal{T}^k \to \mathbb{Q}_{\geq 0}$ such that each tour is routed at a facility, i.e., $|V(T) \cap (F \cup F^k)| = 1$, each client is served, i.e., $\sum_{T \in \mathcal{T}^k : j \in V(T)} x^k(j, T) = d_j^k$ for all $j \in \mathcal{C}$, and the *capacity constraints* $\sum_{j \in \mathcal{C}} x^k(j, T) \leq u$ are satisfied for all $T \in \mathcal{T}^k$. The objective is to minimize the sum of fixed first-stage and expected second-stage costs

$$\sum_{i \in F} f_i + \sum_{k=1}^{m} p_k \cdot \left(\sum_{i \in F^k} f_i^k + \sum_{T \in \mathcal{T}^k} \sum_{e \in E(T)} c^k(e) \right).$$

The remainder of this paper is organized as follows. In Sect. 2, we discuss the complexity of the presented problems and introduce the type of service cost mutability that our local search approach can handle. Afterwards, in Sect. 3, we present our Nested Local Search algorithm for tsUFL, tsUFL-T, and tsUFL-DT and prove its constant approximation guarantees. In Sects. 4 and 5 we construct constant-factor approximations for tsCCFL and tsCLR by applying our local search to instances of the tsUFL-DT problem. Concluding remarks are given in Sect. 6. All omitted proofs can be found in a full version [15].

2   Hardness of Approximation

The tsUFL, tsCCFL, and tsCLR problem generalize the metric UFL problem with uniform demands. So, all hardness results are preserved and these problems are strongly NP-hard. In particular, the inapproximability results of Guha and Khuller [7] and Sviridenko [13] carry over. Hence, there is no 1.463-approximation algorithm for the problems, even when restricted to instances with a fixed metric and service cost 1 and 3, unless P = NP. The tsCLR problem also generalizes the capacitated vehicle routing problem, which is not approximable within a factor less than 1.5, unless P = NP [6]. By a reduction from UFL we obtain the following inapproximability result for tsUFL-T and tsUFL-DT.

Theorem 1. *There is no* 1.463*-factor approximation algorithm for the tsUFL-T and the tsUFL-DT problem, unless P = NP.*

The approximability of the stochastic problems depends on the mutability of the metric, since the hardness result for minimum set cover [5] carries over.

Theorem 2. *For $\varepsilon > 0$, there is no $(1 - \varepsilon) \ln(m)$-approximation algorithm for tsUFL(-T, -DT), tsCCFL, and tsCLR with a general mutable metric, if $P \neq NP$.*

We show in Sect. 3 that the following class of metrics allows constant-factor approximations for the tsUFL, tsUFL-T, and tsUFL-DT problem.

Definition 3. *A family of metrics $(c^k : (\mathcal{C} \cup \mathcal{F})^2 \to \mathbb{Q}_{\geq 0})_{k \in [m]}$ is called order-preserving, if for each facility $i \in \mathcal{F}$ there exists an ordered list of $\mathcal{F} \setminus \{i\}$ that is (simultaneously) non-decreasingly sorted w.r.t. $c^k(i, \cdot)$ for each scenario k.*

Note that order-preserving metrics restrict only the distances among the facilities to form scenario-independent orders. Distances between clients and facilities may vary heavily from one scenario to another. In particular, the closest (open) facility from any client may change from one scenario to another. This generalizes the concept of inflation factors.

3   Nested Local Search Algorithm

In this section we present our Nested Local Search for the tsUFL, tsUFL-T, and tsUFL-DT problem. Given a feasible solution for one of these problems, we say a *feasible move* is an operation that adds an unchosen, deletes a chosen, swaps a chosen with an unchosen facility, or maintains the given facilities, and results in a feasible solution. Speaking of a first-stage or second-stage feasible move, we refer to these operations on first-stage or second-stage facilities, respectively.

Without any bounds on the cost reduction, local search algorithms may have exponential running time. To avoid this, we use the concept of δ-locally optimal solutions. If we guarantee a cost reduction by a factor of $0 < (1 - \delta) < 1$ in each iteration and choose δ appropriately, we prove a polynomial running time.

Definition 4. *A solution is denoted as δ-locally optimal, if no feasible first-stage move linked with any feasible second-stage move in each scenario decreases the total cost by more than a factor $0 < (1 - \delta) < 1$.*

3.1   Algorithm

As the scenarios are linked only to the first stage, we can consider them sequentially, exploring only polynomial many moves in total. Combining all described ideas, we get Nested Local Search illustrated below. The (re-)assignment of the clients to the chosen facilities is done optimally in all solution update steps. We may also assume that the sets of chosen first-stage and second-stage facilities are disjoint. Let `solution` be a feasible solution for one of the problems and denote the total cost by $C(\text{solution})$. We call a feasible first-stage move *unexplored* if this move was not even attempted to apply to `solution`. A feasible second-stage move is called *cost-reducing*, if applying the move does not increase the cost.

Input: Constant $0 < \delta < 1$ and a feasible solution `solution`.
Output: δ-locally optimal solution `solution`.
while *unexplored first-stage move of* `solution` *exists* **do**
    Select unexplored first-stage move, create solution `current`.
    Select most cost-reducing move for each scenario, update `current`.
    while $C(\texttt{current}) \leq (1 - \delta) \cdot C(\texttt{solution})$ **do**
        `solution := current`
        Select most cost-reducing move for each scenario, update `current`.
return `solution`

Nested Local Search

Testing each unexplored move without changing the solution stops the algorithm. Therefore, the algorithm terminates with `solution`. Also, every feasible first-stage move has been evaluated in combination with a most cost-reducing second-stage move for each scenario, but the cost reduction was less than a factor of $(1 - \delta)$. By definition, `solution` thus is δ-locally optimal.

3.2   Analysis

Applying any feasible move to a δ-locally optimal solution does not decrease the cost by more than a factor of $(1 - \delta)$, even if all clients are reassigned optimally afterwards. We use this observation to create new solutions. By comparison of costs we get bounds on the service and the facility cost.

Lemma 5. *Let* C_S, C_S^* *denote the service costs and* C_F, C_F^* *the facility costs of a δ-locally optimal and an arbitrary feasible solution, respectively. Then*

$$C_S - \delta m \cdot |\mathcal{F}| \cdot (C_F + C_S) \leq C_F^* + C_S^*.$$

Lemma 6. *Let* C_S, C_S^* *denote the service costs and* C_F, C_F^* *the facility costs of a δ-locally optimal and an arbitrary feasible solution, respectively. Then*

$$C_F - \delta m \cdot |\mathcal{F}| \cdot (C_F + C_S) \leq C_F^* + 2 \cdot C_S^*.$$

Theorem 7. *Let* $0 < \varepsilon \leq 1$. *Then,* **Nested Local Search** *is a polynomial-time* $(3 + \varepsilon)$-*approximation for tsUFL(-T, -DT) with order-preserving metrics.*

Proof. The number of feasible first-stage and second-stage moves in each scenario is bounded by $|\mathcal{F}|^2 + |\mathcal{F}|$ each. Updating a solution and finding a most cost-reducing move runs in polynomial time. Choosing $\delta := \varepsilon/(8m \cdot |\mathcal{F}|)$ and $0 < \varepsilon \leq 1$ results in a polynomial running time. With Lemmas 5 and 6 we obtain the bound $C_F + C_S \leq 3/(1 - \varepsilon/4) \cdot (C_F^* + C_S^*)$ and the claim follows.

This result is tight, since Arya et al. [1] showed it for the UFL problem.

3.3 Improvements via Cost-Scaling and Greedy Augmentation

The cost-scaling technique introduced by Charikar and Guha [4] can be applied to the problems in a straightforward way (cf. [15]). Applying this technique, we obtain the following strengthened version of Theorem 7.

Theorem 8. *Let $0 < \varepsilon \le 1$. Then,* **Nested Local Search** *with cost-scaling is a $(1 + \sqrt{2} + \varepsilon)$-approximation for tsUFL(-T, -DT) with order-preserving metrics.*

Also, the well-known greedy augmentation technique for facility location problems can be applied in a straightforward way in combination with our Nested Local Search (cf. [15]). Combining all three techniques local search, cost-scaling, and greedy augmentation, we obtain the following stronger result.

Theorem 9. *Let $0 < \varepsilon \le 1$. Then,* **Nested Local Search** *with cost-scaling and greedy augmentation is a $(2.375 + \varepsilon)$-approximation algorithm for the tsUFL, tsUFL-T, and tsUFL-DT problem with order-preserving metrics.*

4 Two-Stage Capacitated-Cable Facility Location

In this section we introduce an approximation algorithm for tsCCFL. Initially, we transform an instance of tsCCFL to an instance of tsUFL-DT and show that the costs of a tsUFL-DT solution can be bounded by the costs of a tsCCFL solution. We then transform a solution to tsUFL-DT to one for tsCCFL.

Lemma 10. *Consider an instance $\mathcal{I}$ of tsCCFL and the instance $\mathcal{J}$ of tsUFL-DT obtained by scaling the demand values with $1/u$, omitting the capacity, and restricting the problem to $G[\mathcal{F} \cup \mathcal{C}]$. Then, for each solution of $\mathcal{I}$ with costs $C_F^* + C_S^*$ there is a solution of $\mathcal{J}$ with costs $C_F' + C_S'$ that $C_F' \le C_F^*$ and $C_S' \le 3 \cdot C_S^*$.*

4.1 Algorithm tsCCFL

We introduce at first an approximation algorithm for the tsCCFL problem with unit demands which we extend to general demand values later. We transform an instance of tsCCFL to an instance of tsUFL-DT as stated in Lemma 10. Then, we apply Nested Local Search with cost-scaling ($\beta = 6.67$) and greedy augmentation, open all obtained facilities and install one unit of capacity on each edge of the obtained trees. If a tree's demand exceeds the capacity we have to relieve this tree. Therefore, we adapt a procedure to relieve overloaded trees used by Ravi and Sinha [11] to approximate a deterministic version of the problem.

In detail, consider each node x where the subtrees of its children have demand at most u and the total demand of the (sub-)tree T_x is greater than u. To relieve overloaded trees, we choose the clients in the subtree of the children of x which are closest to an open facility $F \cup F^k$ and install unit capacity on each edge of the $\lfloor |D_x|/u \rfloor$ closest (w.r.t. c^k) client-facility pairs, but at most one per subtree. Considering one of those client-facility pairs (j_ℓ, i_ℓ) we reroute the

Input: Instance $\mathcal{I}$ of tsCCFL with unit demands and order-preserving metrics.
Output: Approximated solution of the tsCCFL instance $\mathcal{I}$.
Obtain tsUFL-DT instance $\mathcal{J}$ from $\mathcal{I}$ by scaling demand values with $1/u$.
Apply scaling ($\beta = 6.67$), Nested Local Search, and greedy augmentation to $\mathcal{J}$.
Obtain solution $\left(F, F^1, \ldots, F^m, \sigma, \mathcal{T}\right)$ and open all facilities $\left(F, F^1, \ldots, F^m\right)$.
for all *scenarios k* **do**

> Let T_x be the subtree of $T \in \mathcal{T}^k$ rooted at $x \in V(T)$ and $D_x := V(T_x) \cap C$.
> **for all** *facilities $i \in F \cup F^k$* **do**
>
>> Install one copy of the cable on each edge in $E(T_i)$.
>> **while** $|D_i| > u$ **do**
>>
>>> Let $V' := \{x \in V(T_i) \mid |D_x| > u$ and $|D_\ell| \le u$ for each child ℓ of $x\}$.
>>> **for all** $x \in V'$ **do**
>>>
>>>> Let $(j_\ell, i_\ell) := \arg\min_{j' \in D_\ell, i' \in F \cup F^k} c^k(j', i')$ if ℓ is child of x.
>>>> Install one cable on each edge (j_ℓ, i_ℓ) for the $\lfloor |D_x|/u \rfloor$ cheapest pairs (at most one for each child subtree of x).
>>>> Route the whole demand in T_ℓ to i_ℓ via j_ℓ.
>>>> Route remaining demand (in other subtrees T_ℓ of children of x) to a chosen pair or to x such that all new cables are saturated.
>>>> Remove demands in D_i which are satisfied through a new cable.

Remove all cables with flow value zero and all facilities which serve no demand.

Algorithm tsCCFL

demand $|D_\ell| \le u$ of the subtree T_ℓ to the facility i_ℓ. If a newly installed cable is not saturated, this means the demand flow on the arc is less than u, we reroute not satisfied demand of sibling subtrees via x to this facility. We repeat the relieve procedure, until the remaining demand assigned to any x is at most u. In the end, we clean up our solution by removing all unused cables and facilities.

4.2   Analysis

Theorem 11. *Let $\varepsilon > 0$. Then,* ***Algorithm tsCCFL*** *is a $(3.9 + \varepsilon)$-approximation algorithm for tsCCFL with unit demands and order-preserving metrics.*

Proof. First, we show that the solution produced by Algorithm tsCCFL is feasible. Consider a subtree T_i with $|D_i| > u$ in scenario k and let $x \in V'$. We add as many additional cables and reroute demand in subtrees as long as the remaining demand assigned to x is at most u. Hence, V' decreases and therefore $|D_i|$ does. In the end, all edges of the subtrees fulfill the capacity constraint. However, we maybe reroute some demand via a client j_ℓ to a facility i_ℓ. And so we have to ensure that on these paths no capacity constraint is violated. It is maybe the case that after routing demand (via j_ℓ) to i_ℓ and using an arc (j, x), in a further step demand is routed using the arc (x, j). We use flow cancellation to reassign demand flow properly. In particular, flow cancellation only reduces flow in the direction toward the root of a considered tree. If any cable in a scenario k has flow toward the root, its value is, like mentioned before, at most u. Flow away from the root on a cable is only routed once and all the clients in the involved subtree are removed afterwards. The flow value is also at most u, ensuring satisfied cable capacities. The demand routed to a newly installed cable is exactly u. Each client whose demand is assigned to one new cable has a distance to any open facility of at least the length of the new cable. The cost of these cables can be bounded by the cost of the direct connections by aggregating demand. Hence, the total cable cost is bounded by the service costs of the tsUFL-DT solution.

Let C_F^* denote the facility costs and C_S^* the service costs of an optimal solution to an instance of tsCCFL. We know from Lemma 10 that there is a solution to the transformed tsUFL-DT instance with cost $C_F' + C_S'$ such that $C_F' \leq C_F^*$ and $C_S' \leq 3 \cdot C_S^*$. Since our analysis for Nested Local Search permits us to bound the costs by an arbitrary solution, we obtain with Lemmas 5 and 6, rescaling ($\beta = 6.67$), and greedy augmentation a solution with costs

$$C_F + C_S \leq (2 + \ln(6.67) + \varepsilon') \cdot C_F' + \left(1 + \frac{2}{6.67} + \varepsilon'\right) \cdot C_S'$$

$$\leq (3.9 + \varepsilon) \cdot (C_F^* + C_S^*).$$

The best known guarantee for the deterministic version of the problem is $(\rho_{UFL} + \rho_{ST}) \leq 2.88$ [11], with the currently best approximation ratios of Steiner tree [2] and UFL [9]. If we consider the problem spanning only clients with positive demand values, our algorithm ($\beta = 3.33$) yields a $(3.203 + \varepsilon)$-approximation.

4.3   General Demands

Theorem 12. *Let $\varepsilon > 0$. There is a $(6.236 + \varepsilon)$-approximation algorithm for the tsCCFL problem with general demands and order-preserving metrics.*

Proof. The modification of Algorithm tsCCFL to deal with general demand values can be adapted from [11]. In the following we outline briefly the main changes in order to analyze the modifications. Again, we transform the tsCCFL instance as in Lemma 10 and apply rescaling ($\beta = 25.43$), Nested Local Search, and greedy augmentation. For each client which exceeds the capacity ($d_j^k > u$) we install $\lceil d_j^k / u \rceil$ cables on the edge $\{j, \sigma^k(j)\}$ and route its complete demand directly to the facility $\sigma^k(j)$. The service cost for each of these clients can be bounded by twice the costs of their direct connections. The remaining demands are processed as before except that we now accumulate demand to lie in between u and $2u$. Instead of installing one cable, we now install two copies of a cable and route the demand to the corresponding facility. Hence, we now can bound these costs by twice the direct connection costs. Since after greedy augmentation we have $C_S \leq C_S' + C_F'$, we obtain a solution for the tsCCFL problem with general demand values and order-preserving metrics with costs

$$C_F + 2 \cdot C_S \leq (3 + \ln(25.43) + \varepsilon') \cdot C_F' + \left(2 + \frac{2}{25.43} + \varepsilon'\right) \cdot C_S'$$

$$\leq (6.236 + \varepsilon) \cdot (C_F^* + C_S^*).$$

The best guarantee in the deterministic case is $(2\rho_{UFL}+\rho_{ST}) \le 4.37$ [2,9,11]. If we consider the problem spanning only clients with positive demand values, our algorithm ($\beta = 5.572$) yields a $(4.718 + \varepsilon)$-approximation.

5   Two-Stage Capacitated Location Routing

In this section we introduce an approximation algorithm for the tsCLR problem. Initially, we transform a tsCLR instance to one of tsUFL-DT and show that the costs of a tsUFL-DT solution can be bounded by the costs of a tsCLR solution. We then use a solution to tsUFL-DT to build one for tsCLR.

Lemma 13. *Consider an instance $\mathcal{I}$ of tsCLR and the instance $\mathcal{J}$ of tsUFL-DT obtained by scaling the demand values with $2/u$ and omitting the vehicle capacity. Then, for each solution of $\mathcal{I}$ with costs $C_F^* + C_S^*$ there exists a solution of $\mathcal{J}$ with costs $C_F' + C_S'$ such that $C_F' \le C_F^*$ and $C_S' \le 2 \cdot C_S^*$.*

5.1   Algorithm

We introduce an approximation algorithm for tsCLR by using our Nested Local Search with scaling ($\beta = 5.572$) and greedy augmentation on the tsUFL-DT instance obtained by the transformation described in Lemma 13. Consider a tree T_i with demand value D_i routed at facility $i \in F \cup F^k$. If the total demand of the tree satisfies the capacity constraint we obtain a feasible tour by doubling the edges and short-cutting. Otherwise, we relieve the tree by adapting a procedure by Harks et al. [8] for approximating a deterministic version of the problem.

In more detail, we open all obtained facilities. For each client j with demand value at least u we create $\lceil d_j^k/u \rceil$ times the tour $(\sigma^k(j), j, \sigma^k(j))$. Consider a node v where each children's subtree has demand at most u and the total demand of the tree T_v is greater than u. Find a partition $I = I_0 \dot{\cup} \ldots \dot{\cup} I_q$ of the children's subtrees such that the trees of each part obey the capacity constraint and all parts except I_0 have total demand greater than $u/2$. Note that the (sub-)tree structures remain unchanged while generating the partition. Such a partition can be found by a greedy algorithm. Consider a part I_p ($p \ge 1$) and let j be the client in I_p with the smallest distance to an open facility. We construct a tour by doubling the edge $\{\sigma^k(j), j\}$ and all edges contained in I_p and short-cutting. In the end there is only part I_0 with total demand at most u. Again, we create a tour by doubling the edges and short-cutting. Finally, we remove unused facilities to save costs.

5.2   Analysis

Theorem 14. *Let $\varepsilon > 0$. Then, **Algorithm tsCLR** is a $(4.718+\varepsilon)$-approximation algorithm for the tsCLR problem with order-preserving metrics.*

Input: Instance $\mathcal{I}$ of the tsCLR problem.
Output: Approximated solution of the tsCLR instance $\mathcal{I}$.
Obtain tsUFL-DT instance $\mathcal{J}$ from $\mathcal{I}$ by scaling demand values with $2/u$.
Apply scaling ($\beta = 5.572$), Nested Local Search, and greedy augmentation to $\mathcal{J}$.
Obtain solution $(F, F^1, \ldots, F^m, \sigma, \mathcal{T})$ and open all facilities $(F, F^1, \ldots, F^m)$.
for all *scenarios* k **do**
$\quad$ **for all** $j \in C$ with $d_j^k \geq u$ **do**
$\quad\quad$ $\lfloor$ Add $\lceil d_j^k/u \rceil$ copies of the tour $(\sigma^k(j), j, \sigma^k(j))$ and remove d_j^k.
$\quad$ Let T_x be the subtree of $T \in \mathcal{T}^k$ rooted at x, and $D_x := \sum_{j \in C \cap V(T_x)} d_j^k$.
$\quad$ **for all** *facilities* $i \in F \cup F^k$ **do**
$\quad\quad$ **while** $D_i > u$ **do**
$\quad\quad\quad$ Let $v \in \{x \in V(T_i) \mid D_x > u, \ D_\ell \leq u \text{ for all children } \ell \text{ of } x\}$.
$\quad\quad\quad$ Let $I = \{V(T_\ell) \mid \ell \text{ is child of } v\} \cup \{\{v\}\}$.
$\quad\quad\quad$ Find a partition of the trees $I = I_0 \dot\cup \ldots \dot\cup I_q$ such that
$\quad\quad\quad\quad$ $\sum_{x \in I_p} d_x^k \leq u$ for all $p \in \{0, \ldots, q\}$ and
$\quad\quad\quad\quad$ $\sum_{x \in I_p} d_x^k > u/2$ for all $p \in \{1, \ldots, q\}$.
$\quad\quad\quad$ **for all** $p \in \{1, \ldots, q\}$ **do**
$\quad\quad\quad\quad$ Let $(i_\ell, x_\ell) := \arg\min_{i' \in F \cup F^k, x' \in V(I_p)} c^k(i', x')$.
$\quad\quad\quad\quad$ Construct a tour containing all clients in I_p and facility i_ℓ by
$\quad\quad\quad\quad\quad$ doubling (i_ℓ, x_ℓ) and edges of all trees in I_p and short-cutting.
$\quad\quad\quad\quad$ $\lfloor$ Add the tour to the solution and remove corresponding subtrees.
$\quad\quad$ Construct a tour from T_i by doubling all edges and short-cutting.
$\quad\quad$ $\lfloor$ Add the tour to the solution.
Remove all facilities that are not contained in any tour.

Algorithm tsCLR

Proof. For all clients j with demand value $d_j^k \geq u$ in some scenario k we add $\lceil d_j^k/u \rceil$ copies of the tour $(\sigma^k(j), j, \sigma^k(j))$. Such a tour containing client j in scenario k has costs of at most $p_k \cdot \lceil d_j^k/u \rceil \cdot 2 \cdot c^k(\sigma^k(j), j)$. Since $\lceil d_j^k/u \rceil$ is bounded by $2 \cdot d_j^k/u$ for $d_j^k \geq u$, the costs for these clients are bounded by twice the direct connection costs of of these clients.

Consider a tour $T \in \mathcal{T}^k$ in scenario k containing facility i_ℓ and clients in I_p. The costs for T are at most $2 \cdot c^k(i_\ell, x_\ell)$ plus twice the costs of the corresponding subtrees. Since the choice of (i_ℓ, x_ℓ) was minimal w.r.t. c^k and the whole demand in T is at least $u/2$ we obtain $\sum_{x \in V(T)} 2 \cdot d_x^k/u \cdot c^k(\sigma^k(x), x) \geq c^k(i_\ell, x_\ell) \cdot \sum_{x \in V(T)} 2 \cdot d_x^k/u \geq c^k(i_\ell, x_\ell)$. Hence, the clients, carried by such tours, contribute to the costs with at most twice their direct connection costs and twice the costs of the corresponding subtrees. All other tours are built by doubling the edges of corresponding subtrees and short-cutting. These tours contribute to the costs with at most twice the costs of the corresponding subtrees. Summation over all scenarios and clients shows that the tour costs are bounded by twice the direct and twice the tree-connection costs in the constructed solution.

Let C_F^* denote the facility costs and C_S^* the service costs of an optimal solution to an instance of tsCLR. We know with Lemma 13 that there is a solution to the transformed tsUFL-DT instance with costs $C_F' + C_S'$ such that $C_F' \leq C_F^*$ and $C_S' \leq 2 \cdot C_S^*$. Since our analysis for Nested Local Search permits us to bound the costs by an arbitrary solution and $C_S \leq C_S' + C_F'$ holds after greedy augmentation, we obtain with cost-scaling ($\beta = 5.572$), Lemmas 5 and 6, and greedy augmentation a solution with costs

$$C_F + 2 \cdot C_S \leq (3 + \ln(5.572) + \varepsilon') \cdot C_F' + \left(2 + \frac{2}{5.572} + \varepsilon'\right) \cdot C_S'$$

$$\leq (4.718 + \varepsilon) \cdot (C_F^* + C_S^*).$$

The best known approximation algorithm for the deterministic problem has a guarantee of 4.38 and is due to Harks et al. [8]. So our algorithm produces only a slightly worse approximation factor in the two-stage stochastic case.

6   Conclusion

In this paper we introduced Nested Local Search, showing that pure local search applies to metric two-stage stochastic facility location problems. Our analysis lead to a tight $(3 + \varepsilon)$-approximation for the pure local search and to a $(2.375 + \varepsilon)$-factor approximation algorithm for local search combined with rescaling and greedy augmentation techniques. Moreover Nested Local Search allows us to generalize the mutability of the metric in contrast to previous algorithms, which only permit scenario-dependent inflation factors, to order-preserving metrics. Furthermore, we obtained the first constant-factor approximation algorithms for tsCCFL and tsCLR with guarantees $(6.236 + \varepsilon)$ and $(4.718 + \varepsilon)$, respectively.

It would be interesting to know if our new approach combining direct and tree-connections in one facility location problem could lead to improved approximation ratios also for the deterministic problems. Moreover, it would be interesting to study local search techniques for variants of two-stage stochastic capacitated facility location problems, as they proved to be very useful in the deterministic case.

References

1. Arya, V., Garg, N., Khandekar, R., Meyerson, A., Munagala, K., Pandit, V.: Local search heuristics for k-median and facility location problems. SIAM J. Comput. **33**, 544–562 (2004)
2. Byrka, J., Grandoni, F., Rothvoss, T., Sanità, L.: Steiner tree approximation via iterative randomized rounding. J. ACM **60**, 6 (2013)
3. Charikar, M., Chekuri, C., Pál, M.: Sampling bounds for stochastic optimization. In: Chekuri, C., Jansen, K., Rolim, J.D.P., Trevisan, L. (eds.) APPROX/RANDOM -2005. LNCS, vol. 3624, pp. 257–269. Springer, Heidelberg (2005). doi:10.1007/11538462_22

4. Charikar, M., Guha, S.: Improved combinatorial algorithms for the facility location and k-median problems. In: 40th Annual Symposium on Foundations of Computer Science (1999)
5. Dinur, I., Steurer, D.: Analytical approach to parallel repetition. In: Proceedings of the 46th Annual ACM Symposium on Theory of Computing, STOC 2014 (2014)
6. Golden, B.L., Wong, R.T.: Capacitated arc routing problems. Networks **11**, 305–315 (1981)
7. Guha, S., Khuller, S.: Greedy strikes back: improved facility location algorithms. In: Proceedings of the Ninth Annual ACM-SIAM Symposium on Discrete Algorithms (1998)
8. Harks, T., König, F.G., Matuschke, J.: Approximation algorithms for capacitated location routing. Transp. Sci. **47**, 3–22 (2013)
9. Li, S.: A 1.488 approximation algorithm for the uncapacitated facility location problem. In: Aceto, L., Henzinger, M., Sgall, J. (eds.) ICALP 2011. LNCS, vol. 6756, pp. 77–88. Springer, Heidelberg (2011). doi:10.1007/978-3-642-22012-8_5
10. Ravi, R., Sinha, A.: Hedging uncertainty: approximation algorithms for stochastic optimization problems. In: Bienstock, D., Nemhauser, G. (eds.) IPCO 2004. LNCS, vol. 3064, pp. 101–115. Springer, Heidelberg (2004). doi:10.1007/978-3-540-25960-2_8
11. Ravi, R., Sinha, A.: Approximation algorithms for problems combining facility location and network design. Oper. Res. **54**, 73–81 (2006)
12. Shmoys, D.B., Tardos, E., Aardal, K.: Approximation algorithms for facility location problems (extended abstract). In: Proceedings of the Twenty-Ninth Annual ACM Symposium on Theory of Computing (1997)
13. Sviridenko, M.: Unpublished (cf. [14])
14. Shmoys, D.B.: Approximation algorithms for facility location problems. In: Jansen, K., Khuller, S. (eds.) APPROX 2000. LNCS, vol. 1913, pp. 27–32. Springer, Heidelberg (2000). doi:10.1007/3-540-44436-X_4
15. Willamowski, F., Bley, A.: Local search based approximation algorithms for two-stage stochastic location problems. Technical report 2016–036, Operations Research, RWTH Aachen University, October 2016
16. Ye, Y., Zhang, J.: An approximation algorithm for the dynamic facility location problem. In: Cheng, M.X., Li, Y., Du, D.-Z. (eds.) Combinatorial Optimization in Communication Networks. Combinatorial Optimization, vol. 18. Springer, Heidelberg (2006)

Author Index

Printed in the United States
By Bookmasters